Is Mars Habitable?

by Alfred Russel Wallace

PREFACE.

This small volume was commenced as a review article on Professor Percival Lowell's book, Mars and its Canals, with the object of showing that the large amount of new and interesting facts contained in this work did not invalidate the conclusion I had reached in 1902, and stated in my book on Man's Place in the Universe, that Mars was not habitable.

But the more complete presentation of the opposite view in the volume now under discussion required a more detailed examination of the various physical problems involved, and as the subject is one of great, popular, as well as scientific interest, I determined to undertake the task.

This was rendered the more necessary by the fact that in July last Professor Lowell published in the Philosophical Magazine an elaborate mathematical article claiming to demonstrate that, notwithstanding its much greater distance from the sun and its excessively thin atmosphere, Mars possessed a climate on the average equal to that of the south of England, and in its polar and sub-polar regions even less severe than that of the earth. Such a contention of course required to be dealt with, and led me to collect information bearing upon temperature in all its aspects, and so enlarging my criticism that I saw it would be necessary to issue it in book form.

Two of my mathematical friends have pointed out the chief omission which vitiates Professor Lowell's mathematical conclusions--that of a failure to recognise the very large conservative and cumulative effect of a dense atmosphere. This very point however I had already myself discussed in Chapter VI., and by means of some remarkable researches on the heat of the moon and an investigation of the causes of its very low temperature, I have, I think, demonstrated the incorrectness of Mr. Lowell's results. In my last chapter, in which I briefly summarise the whole argument, I have further strengthened the case for very severe cold in Mars, by adducing the rapid lowering of temperature universally caused by diminution of atmospheric pressure, as manifested in the well-known phenomenon of temperate climates at moderate heights even close to the equator, cold climates at greater heights even on extensive plateaux, culminating in arctic climates and perpetual snow at heights where the air is still far denser than it is on the surface of Mars. This argument itself is, in my opinion, conclusive; but it is enforced by two others equally complete, neither of which is adequately met by Mr. Lowell.

The careful examination which I have been led to give to the whole of the phenomena which Mars presents, and especially to the discoveries of Mr. Lowell, has led me to what I hope will be considered a satisfactory physical explanation of them. This explanation, which occupies the whole of my seventh chapter, is founded upon a special mode of origin for Mars, derived from the Meteoritic Hypothesis, now very widely adopted by astronomers and physicists. Then, by a comparison with certain well-known and widely spread geological phenomena, I show how the great features of Mars--the 'canals' and 'oases'--may have been caused. This chapter will perhaps be the most interesting to the general reader, as furnishing a quite natural explanation of features of the planet which have been termed 'non-natural' by Mr. Lowell.

Incidentally, also, I have been led to an explanation of the highly volcanic nature of the moon's surface. This seems to me absolutely to require some such origin as Sir George Darwin has given it, and thus furnishes corroborative proof of the accuracy of the hypothesis that our moon has had an unique origin among the known satellites, in having been thrown off from the earth itself.

I am indebted to Professor J. H. Poynting, of the University of Birmingham, for valuable suggestions on some of the more difficult points of mathematical physics here discussed, and also for the critical note (at the end of Chapter V.) on Professor Lowell's estimate of the temperature of Mars.

BROADSTONE, DORSET, October 1907.

TABLE OF CONTENTS.

CHAPTER I.

CHAPTER II.

CHAPTER III.

CHAPTER IV.

the canals.

CHAPTER VIII.

PAGE SUMMARY AND CONCLUSION, --The canals the origin of Mr. Lowell's theory --Best explained as natural features --Evaporation difficulty not met by Mr. Lowell --How did Martians live without the canals --Radiation due to scanty atmosphere not taken account of --Three independent proofs of low temperature and uninhabitability of Mars --Conclusion.

CHAPTER I.

EARLY OBSERVERS OF MARS.

Few persons except astronomers fully realise that of all the planets of the Solar system the only one whose solid surface has been seen with certainty is Mars; and, very fortunately, that is also the only one which is sufficiently near to us for the physical features of the surface to be determined with any accuracy, even if we could see it in the other planets. Of Venus we probably see only the upper surface of its cloudy atmosphere.[1] As regards Jupiter and Saturn this is still more certain, since their low density will only permit of a comparatively small proportion of their huge bulk being solid. Their belts are but the cloud-strata of their upper atmosphere, perhaps thousands of miles above their solid surfaces, and a somewhat similar condition seems to prevail in the far more remote planets Uranus and Neptune. It has thus happened, that, although as telescopic objects of interest and beauty, the marvellous rings of Saturn, the belts and ever-changing aspects of the satellites of Jupiter, and the moon-like phases of Venus, together with its extreme brilliancy, still remain unsurpassed, yet the greater amount of details of these features when examined with the powerful instruments of the nineteenth century have neither added much to our knowledge of the planets themselves or led to any sensational theories calculated to attract the popular imagination.

[Footnote 1: Mercury also seems to have a scanty atmosphere, but as its mass is only one-thirtieth that of the earth it can retain only the heavier gases, and its atmosphere may be dust-laden, as is that of Mars, according to Mr. Lowell. Its dusky markings, as seen by Schiaparelli, seem to be permanent, and they are also for considerable periods unchangeable in position, indicating that the planet keeps the same face towards the sun as does Venus. This was confirmed by Mr. Lowell in 1896. Its distance from us and unfavourable position for observation must prevent us from obtaining any detailed knowledge of its actual surface, though its low reflective power indicates that the surface may be really visible.]

But in the case of Mars the progress of discovery has had a very different result. The most obvious peculiarity of this planet--its polar snow-caps--were seen about 250 years ago, but they were first proved to increase and decrease alternately, in the summer and winter of each hemisphere, by Sir William Herschell in the latter part of the eighteenth century. This fact gave the impulse to that idea of similarity in the conditions of Mars and the earth, which the recognition of many large dusky patches and streaks as water, and the more ruddy and brighter portions as land, further increased. Added to this, a day only about half an hour longer than our own, and a succession of seasons of the same character as ours but of nearly double the length owing to its much longer year, seemed to leave little wanting to render this planet a true earth on a smaller scale. It was therefore very natural to suppose that it must be inhabited, and that we should some day obtain evidence of the fact.

The Canals discovered by Schiaparelli.

Hence the great interest excited when Schiaparelli, at the Milan Observatory, during the very favourable opposition of 1877 and 1879, observed that the whole of the tropical and temperate regions from 60?N. to 60?S. Lat. were covered with a remarkable network of broader curved and narrower straight lines of a dark colour. At each successive favourable opposition, these strange objects called canali (channels) by their discoverer, but rather misleadingly 'canals' in England and America, were observed by means of all the great telescopes in the world, and their reality and general features became well established. In Schiaparelli's first map they were represented as being much broader and less sharply defined than he himself and other observers found by later and equally favourable observations that they really were.

Discovery of the Double Canals.

In 1881 another strange feature was discovered by Schiaparelli, who found that about twenty canals which had previously been seen single were now distinctly double, that is, that they consisted of two parallel lines, equally distinct and either very close together or a considerable distance apart. This curious appearance was at first thought to be due to some instrumental defect or optical illusion; but as it was soon confirmed by other observers with the best instruments and in widely different localities it became in time accepted as a real phenomenon of the planet's surface.

Round Spots discovered in 1892.

At the favourable opposition of 1892, Mr. W. H. Pickering noticed that besides the 'seas' of various sizes there were numerous very small black spots apparently quite circular and occurring at every intersection or starting-point of the 'canals.' Many of these had been seen by Schiaparelli as larger and ill-defined dark patches, and were termed seas or lakes; but Mr. Pickering's observatory was at Arequipa in Peru, about 8000 feet above the sea, and with such perfect atmospheric conditions as were, in his opinion, equal to a doubling of telescopic aperture. They were soon detected by other observers, especially by Mr. Lowell in 1894, who thus wrote of them:

"Scattered over the orange-ochre groundwork of the continental regions of the planet, are any number of dark round spots. How many there may be it is not possible to state, as the better the seeing, the more of them there seem to be. In spite, however, of their great number, there is no instance of one unconnected with a canal. What is more, there is apparently none that does not lie at the junction of several canals. Reversely, all the junctions appear to be provided with spots. Plotted upon a globe they and their connecting canals make a most curious network over all the orange-ochre equatorial parts of the planet, a mass of lines and knots, the one marking being as omnipresent as the other."

Changes of Colour recognised.

During the oppositions of 1892 and 1894 it was fully recognised that a regular course of change occurred dependent upon the succession of the seasons, as had been first suggested by Schiaparelli. As the polar snows melt the adjacent seas appear to overflow and spread out as far as the tropics, and are often seen to assume a distinctly green colour. These remarkable changes and the extraordinary phenomena of perfect straight lines crossing each other over a large portion of the planet's surface, with the circular spots at their intersections, had such an appearance of artificiality that the idea that they were really 'canals' made by intelligent beings for purposes of irrigation, was first hinted at, and then adopted as the only intelligible explanation, by Mr. Lowell and a few other persons. This at once seized upon the public imagination and was spread by the newspapers and magazines over the whole civilised world.

Existence of Seas doubted.

At this time (1894) it began to be doubted whether there were any seas at all on Mars. Professor Pickering thought they were far more limited in size than had been supposed, and even might not exist as true seas. Professor Barnard, with the Lick thirty-six inch telescope, discerned an astonishing wealth of detail on the surface of Mars, so intricate, minute, and abundant, that it baffled all attempts to delineate it; and these peculiarities were seen upon the supposed seas as well as on the land-surfaces. In fact, under the best conditions these 'seas' lost all trace of uniformity, their appearance being that of a mountainous country, broken by ridges, rifts, and canyons, seen from a great elevation. As we shall see later on these doubts soon became certainties, and it is now almost universally admitted that Mars possesses no permanent bodies of water.

CHAPTER II.

MR. PERCIVAL LOWELL'S DISCOVERIES AND THEORIES.

The Observatory in Arizona.

In 1894, after a careful search for the best atmospheric conditions, Mr. Lowell established his observatory near the town of Flagstaff in Arizona, in a very dry and uniform climate, and at an elevation of 7300 feet above the sea. He then possessed a fine equatorial telescope of 18 inches aperture and 26 feet focal length, besides two smaller ones, all of the best quality. To these he added in 1896 a telescope with 24 inch object glass, the last work of the celebrated firm of Alvan Clark & Sons, with which he has made his later discoveries. He thus became perhaps more favourably situated than any astronomer in the northern hemisphere, and during the last twelve years has made a specialty of the study of Mars, besides doing much valuable astronomical work on other planets.

Mr, Lowell's recent Books upon Mars.

In 1905 Mr. Lowell published an illustrated volume giving a full account of his observations of Mars from 1894 to 1903, chiefly for the use of astronomers; and he has now given us a popular volume summarising the whole of his work on the planet, and published both in America and England by the Macmillan Company. This very interesting volume is fully illustrated with twenty plates, four of them coloured, and more than forty figures in the text, showing the great variety of details from which the larger general maps have been constructed.

Non-natural Features of Mars.

But what renders this work especially interesting to all intelligent readers is, that the author has here, for the first time, fully set forth his views both as to the habitability of Mars and as to its being actually inhabited by beings comparable with ourselves in intellect. The larger part of the work is in fact devoted to a detailed description of what he terms the 'Non-natural Features' of the planet's surface, including especially a full account of the 'Canals,' single and double; the 'Oases,' as he terms the dark spots at their intersections; and the varying visibility of both, depending partly on the Martian seasons; while the five concluding chapters deal with the possibility of animal life and the evidence in favour of it. He also upholds the theory of the canals having been constructed for the purpose of 'husbanding' the scanty water-supply that exists; and throughout the whole of this argument he clearly shows that he considers the evidence to be satisfactory, and that the only intelligible explanation of the whole of the phenomena he so clearly sets forth is, that the inhabitants of Mars have carried out on their small and naturally inhospitable planet a vast system of irrigation-works, far greater both in its extent, in its utility, and its effect upon their world as a habitation for civilised beings, than anything we have yet done upon our earth, where our destructive agencies are perhaps more prominent than those of an improving and recuperative character.

A Challenge to the Thinking World.

This volume is therefore in the nature of a challenge, not so much to astronomers as to the educated world at large, to investigate the evidence for so portentous a conclusion. To do this requires only a general acquaintance with modern science, more especially with mechanics and physics, while the main contention (with which I shall chiefly deal) that the features termed 'canals' are really works of art and necessitate the presence of intelligent organic beings, requires only care and judgment in drawing conclusions from admitted facts. As I have already paid some attention to this problem and have expressed the opinion that Mars is not habitable,[2] judging from the evidence then available, and as few men of science have the leisure required for a careful examination of so speculative a subject, I propose here to point out what the facts, as stated by Mr. Lowell himself, do not render even probable much less prove. Incidentally, I may be able to adduce evidence of a more or less weighty character, which seems to negative the possibility of any high form of animal life on Mars, and, a fortiori, the development of such life as might culminate in a being equal or superior to ourselves. As most popular works on Astronomy for the last ten years at least, as well as many scientific periodicals and popular magazines, have reproduced some of the maps of Mars by Schiaparelli, Lowell, and others, the general appearance of its surface will be familiar to most readers, who will thus be fully able to appreciate Mr. Lowell's account of his own further discoveries which I may have to quote. One of the best of these maps I am able to give as a frontispiece to this volume, and to this I shall mainly refer.

[Footnote 2: Man's Place in the Universe p. 267 (1903).]

The Canals as described by Mr. Lowell.

In the clear atmosphere of Arizona, Mr. Lowell has been able on various favourable occasions to detect a network of straight lines, meeting or crossing each other at various angles, and often extending to a thousand or even over two thousand miles in length. They are seen to cross both the light and the dark regions of the planet's surface, often extending up to or starting from the polar snow-caps. Most of these lines are so fine as only to be visible on special occasions of atmospheric clearness and steadiness, which hardly ever occur at lowland stations, even with the best instruments, and almost all are seen to be as perfectly straight as if drawn with a ruler.

The Double Canals.

Under exceptionally favourable conditions, many of the lines that have been already seen single appear double--a pair of equally fine lines exactly parallel throughout their whole length, and appearing, as Mr. Lowell says, "clear cut upon the disc, its twin lines like the rails of a railway track." Both Schiaparelli and Lowell were at first so surprised at this phenomenon that they thought it must be an optical illusion, and it was only after many observations in different years, and by the application of every conceivable test, that they both became convinced that they witnessed a real feature of the planet's surface. Mr. Lowell says he has now seen them hundreds of times, and that his first view of one was 'the most startlingly impressive' sight he has ever witnessed.

Dimensions of the Canals.

A few dimensions of these strange objects must be given in order that readers may appreciate their full strangeness and inexplicability. Out of more than four hundred canals seen and recorded by Mr. Lowell, fifty-one, or about one eighth, are either constantly or occasionally seen to be double, the appearance of duplicity being more or less periodical. Of 'canals' generally, Mr. Lowell states that they vary in length from a few hundred to a few thousand miles long, one of the largest being the Phison, which he terms 'a typical double canal,' and which is said to be 2250 miles long, while the distance between its two constituents is about 130 miles.[3] The actual width of each canal is from a minimum of about a mile up to several miles, in one case over twenty. A great feature of the doubles is, that they are strictly parallel throughout their whole course, and that in almost all cases they are so truly straight as to form parts of a great circle of the planet's sphere. A few however follow a gradual but very distinct curve, and such of these as are double present the same strict parallelism as those which are straight.

[Footnote 3: This is on the opposite side of Mars from that shown in the frontispiece.]

Canals extend across the Seas.

It was only after seventeen years of observation of the canals that it was found that they extended also into and across the dark spots and surfaces which by the earlier observers were termed seas, and which then formed the only clearly distinguishable and permanent marks on the planet's surface. At the present time, Professor Lowell states that this "curious triangulation has been traced over almost every portion of the planet's surface, whether dark or light, whether greenish, ochre, or brown in colour." In some parts they are much closer together than in others, "forming a perfect network of lines and spots, so that to identify them all was a matter of extreme difficulty." Two such portions are figured at pages 247 and 256 of Mr. Lowell's volume.

The Oases.

The curious circular black spots which are seen at the intersections of many of the canals, and which in some parts of the surface are very numerous, are said to be more difficult of detection than even the lines, being often blurred or rendered completely invisible by slight irregularities in our own atmosphere, while the canals themselves continue visible. About 180 of these have now been found, and the more prominent of them are estimated to vary from 75 to 100 miles in diameter. There are however many much smaller, down to minute and barely visible black points. Yet they all seem a little larger than the canals which enter them. Where the canals are double, the spots (or 'oases' as Mr. Lowell terms them) lie between the two parallel canals.

No one can read this book without admiration for the extreme perseverance in long continued and successful observation, the results of which are here recorded; and I myself accept unreservedly the substantial accuracy of the whole series. It must however always be remembered that the growth of knowledge of the detailed markings has been very gradual, and that much of it has only been seen under very rare and exceptional conditions. It is therefore quite possible that, if at some future time a further considerable advance in instrumental power should be made, or a still more favourable locality be found, the new discoveries might so modify present appearances as to render a satisfactory explanation of them more easy than it is at present.

But though I wish to do the fullest justice to Mr. Lowell's technical skill and long years of persevering work, which have brought to light the most complex and remarkable appearances that any of the heavenly bodies present to us, I am obliged absolutely to part company with him as regards the startling theory of artificial production which he thinks alone adequate to explain them. So much is this the case, that the very phenomena, which to him seem to demonstrate the intervention of intelligent beings working for the improvement of their own environment, are those which seem to me to bear the unmistakable impress of being due to natural forces, while they are wholly unintelligible as being useful works of art. I refer of course to the great system of what are termed 'canals,' whether single or double. Of these I shall give my own interpretation later on.

CHAPTER III.

THE CLIMATE AND PHYSIOGRAPHY OF MARS.

Mr. Lowell admits, and indeed urges strongly, that there are no permanent bodies of water on Mars; that the dark spaces and spots, thought by the early observers to be seas, are certainly not so now, though they may have been at an earlier period; that true clouds are rare, even if they exist, the appearances that have been taken for them being either dust-storms or a surface haze; that there is consequently no rain, and that large portions (about two-thirds) of the planet's surface have all the characteristics of desert regions.

Snow-caps the only Source of Water.

This state of things is supposed to be ameliorated by the fact of the polar snows, which in the winter cover the arctic and about half the temperate regions of each hemisphere alternately. The maximum of the northern snow-caps is reached at a period of the Martian winter corresponding to the end of February with us. About the end of March the cap begins to shrink in size (in the Northern Hemisphere), and this goes on so rapidly that early in the June of Mars it is reduced to its minimum. About the same time changes of colour take place in the adjacent darker portions of the surface, which become at first bluish, and later a decided blue-green; but by far the larger portion, including almost all the equatorial regions of the planet, remain always of a reddish-ochre tint.[4]

[Footnote 4: In 1890 at Mount Wilson, California, Mr. W.H. Pickering's photographs of Mars on April 9th showed the southern polar cap of moderate dimensions, but with a large dim adjacent area. Twenty-four hours later a corresponding plate showed this same area brilliantly white; the result apparently of a great Martian snowfall. In 1882 the same observer witnessed the steady disappearance of 1,600,000 square miles of the southern snow-cap, an area nearly one-third of that hemisphere of the planet.]

The rapid and comparatively early disappearance of the white covering is, very reasonably, supposed to prove that it is of small thickness, corresponding perhaps to about a foot or two of snow in north-temperate America and Europe, and that by the increasing amount of sun-heat it is converted, partly into liquid and partly into vapour. Coincident with this disappearance and as a presumed result of the water (or other liquid) producing inundations, the bluish-green tinge which appears on the previously dark portion of the surface is supposed to be due to a rapid growth of vegetation.

But the evidence on this point does not seem to be clear or harmonious, for in the four coloured plates showing the planet's surface at successive Martian dates from December 30th to February 21st, not only is a considerable extent of the south temperate zone shown to change rapidly from bluish-green to chocolate-brown and then again to bluish-green, but the portions furthest from the supposed fertilising overflow are permanently green, as are also considerable portions in the opposite or northern hemisphere, which one would think would then be completely dried up.

No Hills upon Mars.

The special point to which I here wish to call attention is this. Mr. Lowell's main contention is, that the surface of Mars is wonderfully smooth and level. Not only are there no mountains, but there are no hills or valleys or plateaux. This assumption is absolutely essential to support the other great assumption, that the wonderful network of perfectly straight lines over nearly the whole surface of the planet are irrigation canals. It is not alleged that irregularities or undulations of a few hundreds or even one or two thousands of feet could possibly be detected, while certainly all we know of planetary formation or structure point strongly towards some inequalities of surface. Mr. Lowell admits that the dark portions of the surface, when examined on the terminator (the margin of the illuminated portion), do look like hollows and may be the beds of dried-up seas; yet the supposed canals run across these old sea-beds in perfect straight lines just as they do across the many thousand miles of what are admitted to be deserts--which he describes in these forcible terms: "Pitiless as our deserts are, they are but faint forecasts of the state of things existent on Mars at the present time."

It appears, then, that Mr. Lowell has to face this dilemma--Only if the whole surface of Mars is an almost perfect level could the enormous network of straight canals, each from hundreds to thousands of miles long, have been possibly constructed by intelligent beings for purposes of irrigation; but, if a complete and universal level surface exists no such system would be necessary. For on a level surface--or on a surface slightly inclined from the poles towards the equator, which would be advantageous in either case--the melting water would of itself spread over the ground and naturally irrigate as much of the surface as it was possible for it to reach. If the surface were not level, but consisted of slight elevations and expressions to the extent of a few scores or a few hundreds of feet, then there would be no possible advantage in cutting straight troughs through these elevations in various directions with water flowing at the bottom of them. In neither case, and in hardly any conceivable case, could these perfectly straight canals, cutting across each other in every direction and at very varying angles, be of any use, or be the work of an intelligent race, if any such race could possibly have been developed under the adverse conditions which exist in Mars.

The Scanty Water-supply.

But further, if there were any superfluity of water derived from the melting snow beyond what was sufficient to moisten the hollows indicated by the darker portions of the surface, which at the time the water reaches them acquire a green tint (a superfluity under the circumstances highly improbable), that superfluity could be best utilised by widening, however little, the borders to which natural overflow had carried it. Any attempt to make that scanty surplus, by means of overflowing canals, travel across the equator into the opposite hemisphere, through such a terrible desert region and exposed to such a cloudless sky as Mr. Lowell describes, would be the work of a body of madmen rather than of intelligent beings. It may be safely asserted that not one drop of water would escape evaporation or insoak at even a hundred miles from its source. [5]

[Footnote 5: What the evaporation is likely to be in Mars may be estimated by the fact, stated by Professor J.W. Gregory in his recent volume on 'Australia' in Stanford's Compendium, that in North-West Victoria evaporation is at the rate of ten feet per annum, while in Central Australia it is very much more. The greatly diminished atmospheric pressure in Mars will probably more than balance the loss of sun-heat in producing rapid evaporation.]

Miss Clerke on the Scanty Water-supply.

On this point I am supported by no less an authority than the historian of modern astronomy, the late Miss Agnes Clerke. In the Edinburgh Review (of October 1896) there is an article entitled 'New Views about Mars,' exhibiting the writer's characteristic fulness of knowledge and charm of style. Speaking of Mr. Lowell's idea of the 'canals' carrying the surplus water across the equator, far into the opposite hemisphere, for purposes of irrigation there (which we see he again states in the present volume), Miss Clerke writes: "We can hardly imagine so shrewd a people as the irrigators of Thule and Hellas[6] wasting labour, and the life-giving fluid, after so unprofitable a fashion. There is every reason to believe that the Martian snow-caps are quite flimsy structures. Their material might be called snow souffl? since, owing to the small power of gravity on Mars, snow is almost three times lighter there than here. Consequently, its own weight can have very little effect in rendering it compact. Nor, indeed, is there time for much settling down. The calotte does not form until several months after the winter solstice, and it begins to melt, as a rule, shortly after the vernal equinox. (The interval between these two epochs in the southern hemisphere of Mars is 176 days.) The snow lies on the ground, at the outside, a couple of months. At times it melts while it is still fresh fallen. Thus, at the opposition of 1881-82 the spreading of the northern snows was delayed until seven weeks after the equinox: and they had, accordingly, no sooner reached their maximum than they began to decline. And Professor Pickering's photographs of April 9th and 10th, 1890, proved that the southern calotte may assume its definitive proportions in a single night.

[Footnote 6: Areas on Mars so named.]

"No attempt has yet been made to estimate the quantity of water derivable from the melting of one of these formations; yet the experiment is worth trying as a help towards defining ideas. Let us grant that the average depth of snow in them, of the delicate Martian kind, is twenty feet, equivalent at the most to one foot of water. The maximum area covered, of 2,400,000 square miles, is nearly equal to that of the United States, while the whole globe of Mars measures 55,500,000 square miles, of which one-third, on the present hypothesis, is under cultivation, and in need of water. Nearly the whole of the dark areas, as we know, are situated in the southern hemisphere, of which they extend over, at the very least, 17,000,000 square miles; that is to say, they cover an area, in round numbers, seven times that of the snow-cap. Only one-seventh of a foot of water, accordingly, could possibly be made available for their fertilisation, supposing them to get the entire advantage of the spring freshet. Upon a stint of less than two inches of water these fertile lands are expected to flourish and bear abundant crops; and since they completely enclose the polar area they are necessarily served first. The great emissaries for carrying off the surplus of their aqueous riches, would then appear to be superfluous constructions, nor is it likely that the share in those riches due to the canals and oases, intricately dividing up the wide, dry, continental plains, can ever be realised.

"We have assumed, in our little calculation, that the entire contents of a polar hood turn to water; but in actual fact a considerable proportion of them must pass directly into vapour, omitting the intermediate stage. Even with us a large quantity of snow is removed aerially; and in the rare atmosphere of Mars this cause of waste must be especially effective. Thus the polar reservoirs are despoiled in the act of being opened. Further objections might be taken to Mr. Lowell's irrigation scheme, but enough has been said to show that it is hopelessly unworkable."

It will be seen that the writer of this article accepted the existence of water on Mars, on the testimony of Sir W. Huggins, which, in view of later observations, he has himself acknowledged to be valueless. Dr. Johnstone Stoney's proof of its absence, derived from the molecular theory of gases, had not then been made public.

Description of some of the Canals.

At the end of his volume Mr. Lowell gives a large chart of Mars on Mercator's projection, showing the canals and other features seen during the opposition of 1905. This contains many canals not shown on the map here reproduced (see frontispiece), and some of the differences between the two are very puzzling. Looking at our map, which shows the north-polar snow below, so that the south pole is out of the view at the top of the map, the central feature is the large spot Ascraeeus Lucus, from which ten canals diverge centrally, and four from the sides, forming wide double canals, fourteen in all. There is also a canal named Ulysses, which here passes far to the right of the spot, but in the large chart enters it centrally. Looking at our map we see, going downwards a little to the left, the canal Udon, which runs through a dark area quite to the outer margin. In the dark area, however, there is shown on the chart a spot Aspledon Lucus, where five canals meet, and if this is taken as a terminus the Udon canal is almost exactly 2000 miles long, and another on its right, Lapadon, is the same length, while Ich, running in a slightly curved line to a large spot (Lucus Castorius on the chart) is still longer. The Ulysses canal, which (on the chart) runs straight from the point of the Mare Sirenum to the Astraeeus Lucus is about 2200 miles long. Others however are even longer, and Mr. Lowell says: "With them 2000 miles is common; while many exceed 2500; and the Eumenides-Orcus is 3540 miles from the point where it leaves Lucus Phoeniceus to where it enters the Trivium Charontis." This last canal is barely visible on our map, its commencement being indicated by the word Eumenides.

The Trivium Charontis is situated just beyond the right-hand margin of our map. It is a triangular dark area, the sides about 200 miles long, and it is shown on the chart as being the centre from which radiate thirteen canals. Another centre is Aquae Calidae situated at the point of a dark area running obliquely from 55?to 35?N. latitude, and, as shown on a map of the opposite hemisphere to our map, has nearly twenty canals radiating from it in almost every direction. Here at all events there seems to be no special connection with the polar snow-caps, and the radiating lines seem to have no intelligent purpose whatever, but are such as might result from fractures in a glass globe produced by firing at it with very small shots one at a time. Taking the whole series of them, Mr. Lowell very justly compares them to "a network which triangulates the surface of the planet like a geodetic survey, into polygons of all shapes and sizes."

At the very lowest estimate the total length of the canals observed and mapped by Mr. Lowell must be over a hundred thousand miles, while he assures us that numbers of others have been seen over the whole surface, but so faintly or on such rare occasions as to elude all attempts to fix their position with certainty. But these, being of the same character and evidently forming part of the same system, must also be artificial, and thus we are led to a system of irrigation of almost unimaginable magnitude on a planet which has no mountains, no rivers, and no rain to support it; whose whole water-supply is derived from polar snows, the amount of which is ludicrously inadequate to need any such world-wide system; while the low atmospheric pressure would lead to rapid evaporation, thus greatly diminishing the small amount of moisture that is available. Everyone must, I think, agree with Miss Clerke, that, even admitting the assumption that the polar snows consist of frozen water, the excessively scanty amount of water thus obtained would render any scheme of world-wide distribution of it hopelessly unworkable.

The very remarkable phenomena of the duplication of many of the lines, together with the darkspots--the so-called oases--at their intersections, are doubtless all connected in some unknown way with the constitution and past history of the planet; but, on the theory of the whole being works of art, they certainly do not help to remove any of the difficulties which have been shown to attend the theory that the single lines represent artificial canals of irrigation with a strip of verdure on each side of them produced by their overflow.

Lowell on the Purpose of the Canals.

Before leaving this subject it will be well to quote Mr. Lowell's own words as to the supposed perfectly level surface of Mars, and his interpretation of the origin and purpose of the 'canals':

"A body of planetary size, if unrotating, becomes a sphere, except for solar tidal deformation; if rotating, it takes on a spheroidal form exactly expressive, so far as observation goes, of the so-called centrifugal force at work. Mars presents such a figure, being flattened out to correspond to its axial rotation. Its surface therefore is in fluid equilibrium, or, in other words, a particle of liquid at any point of its surface at the present time would stay where it was devoid of inclination to move elsewhere. Now the water which quickens the verdure of the canals moves from the pole down to the equator as the season advances. This it does then irrespective of gravity. No natural force propels it, and the inference is forthright and inevitable that it is artificially helped to its end. There seems to be no escape from this deduction. Water only flows downhill, and there is no such thing as downhill on a surface already in fluid equilibrium. A few canals might presumably be so situated that their flow could, by inequality of terrane, lie equatorward, but not all....Now it is not in particular but by general consent that the canal-system of Mars develops from pole to equator. From the respective times at which the minima take place, it appears that the canal quickening occupies fifty-two days, as evidenced by the successive vegetal darkenings, to descend from latitude 72?north to latitude 0? a journey of 2650 miles. This gives for the water a speed of fifty-one miles a day, or 2.1 miles an hour. The rate of progression is remarkably uniform, and this abets the deduction as to assisted transference. But the fact is more unnatural yet. The growth pays no regard to the equator, but proceeds across it as if it did not exist into the planet's other hemisphere. Here is something still more telling than travel to this point. For even if we suppose, for the sake of argument, that natural forces took the water down to the equator, their action must there be certainly reversed, and the equator prove a dead-line, to pass which were impossible" (pp. 374-5).

I think my readers will agree with me that this whole argument is one of the most curious ever put forth seriously by an eminent man of science. Because the polar compression of Mars is about what calculation shows it ought to be in accordance with its rate of rotation, its surface is in a state of 'fluid equilibrium,' and must therefore be absolutely level throughout. But the polar compression of the earth equally agrees with calculation; therefore its surface is also in 'fluid equilibrium'; therefore it also ought to be as perfectly level on land as it is on the ocean surface! But as we know this is very far from being the case, why must it be so in Mars? Are we to suppose Mars to have been formed in some totally different way from other planets, and that there neither is nor ever has been any reaction between its interior and exterior forces? Again, the assumption of perfect flatness is directly opposed to all observation and all analogy with what we see on the earth and moon. It gives no account whatever of the numerous and large dark patches, once termed seas, but now found to be not so, and to be full of detailed markings and varied depths of shadow. To suppose that these are all the same dead-level as the light-coloured portions are assumed to be, implies that the darkness is one of material and colour only, not of diversified contour, which again is contrary to experience, since difference of material with us always leads to differences in rate of degradation, and hence of diversified contour, as these dark spaces actually show themselves under favourable conditions to independent observers.

Lowell on the System of Canals as a whole.

We will now see what Mr. Lowell claims to be the plain teaching of the 'canals' as a whole:

"But last and all-embracing in its import is the system which the canals form. Instead of running at hap-hazard, the canals are interconnected in a most remarkable manner. They seek centres instead of avoiding them. The centres are linked thus perfectly one with another, an arrangement which could not result from centres, whether of explosion or otherwise, which were themselves discrete. Furthermore, the system covers the whole surface of the planet, dark areas and light ones alike, a world-wide distribution which exceeds the bounds of natural possibility. Any force which could act longitudinally on such a scale must be limited latitudinally in its action, as witness the belts of Jupiter and the spots upon the sun. Rotational, climatic, or other physical cause could not fail of zonal expression. Yet these lines are grandly indifferent to such competing influences. Finally, the system, after meshing the surface in its entirety, runs straight into the polar caps.

"It is, then, a system whose end and aim is the tapping of the snow-cap for the water there semi-annually let loose; then to distribute it over the planet's face" (p. 373).

Here, again, we have curiously weak arguments adduced to support the view that these numerous straight lines imply works of art rather than of nature, especially in the comparison made with the belts of Jupiter and the spots on the sun, both purely atmospheric phenomena, whereas the lines on Mars are on the solid surface of the planet. Why should there be any resemblance between them? Every fact stated in the above quotation, always keeping in mind the physical conditions of the planet--its very tenuous atmosphere and rainless desert-surface--seem wholly in favour of a purely natural as opposed to an artificial origin; and at the close of this discussion I shall suggest one which seems to me to be at least possible, and to explain the whole series of the phenomena set forth and largely discovered by Mr. Lowell, in a simpler and more probable manner than does his tremendous assumption of their being works of art. Readers who may not possess Mr. Lowell's volume will find three of his most recent maps of the 'canals' reproduced in Nature of October 11th, 1906.

CHAPTER IV.

IS ANIMAL LIFE POSSIBLE ON MARS?

Having now shown, that, even admitting the accuracy of all Mr. Lowell's observations, and provisionally accepting all his chief conclusions as to the climate, the nature of the snow-caps, the vegetation, and the animal life of Mars, yet his interpretation of the lines on its surface as being veritably 'canals,' constructed by intelligent beings for the special purpose of carrying water to the more arid regions, is wholly erroneous and rationally inconceivable. I now proceed to discuss his more fundamental position as to the actual habitability of Mars by a highly organised and intellectual race of material organic beings.

Water and Air essential to Life.

Here, fortunately, the issue is rendered very simple, because Mr. Lowell fully recognises the identity of the constitution of matter and of physical laws throughout the solar-system, and that for any high form of organic life certain conditions which are absolutely essential on our earth must also exist in Mars. He admits, for example, that water is essential, that an atmosphere containing oxygen, nitrogen, aqueous vapour, and carbonic acid gas is essential, and that an abundant vegetation is essential; and these of course involve a surface-temperature through a considerable portion of the year that renders the existence of these--especially of water--possible and available for the purposes of a high and abundant animal life.

Blue Colour the only Evidence of Water.

In attempting to show that these essentials actually exist on Mars he is not very successful. He adduces evidence of an atmosphere, but of an exceedingly scanty one, since the greatest amount he can give to it is-- "not more than about four inches of barometric pressure as we reckon it";[7] and he assumes, as he has a fair right to do till disproved, that it consists of oxygen and nitrogen, with carbon-dioxide and water-vapour, in approximately the same proportions as with us. With regard to the last item-- the water-vapour--there are however many serious difficulties. The water-vapour of our atmosphere is derived from the enormous area of our seas, oceans, lakes, and rivers, as well as from the evaporation from heated lands and tropical forests of much of the moisture produced by frequent and abundant rains. All these sources of supply are admittedly absent from Mars, which has no permanent bodies of water, no rain, and tropical regions which are almost entirely desert. Many writers have therefore doubted the existence of water in any form upon this planet, supposing that the snow-caps are not formed of frozen water but of carbon-dioxide, or some other heavy gas, in a frozen state; and Mr. Lowell evidently feels this to be a difficulty, since the only fact he is able to adduce in favour of the melting snows of the polar caps producing water is, that at the time they are melting a marginal blue band appears which accompanies them in their retreat, and this blue colour is said to prove conclusively that the liquid is not carbonic acid but water. This point he dwells upon repeatedly, stating, of these blue borders: "This excludes the possibility of their being formed by carbon-dioxide, and shows that of all the substances we know the material composing them must be water."

[Footnote 7: In a paper written since the book appeared the density of air at the surface of Mars is said to be 1/12 of the earth's.]

This is the only proof of the existence of water he adduces, and it is certainly a most extraordinary and futile one. For it is perfectly well known that although water, in large masses and by transmitted light, is of a blue colour, yet shallow water by reflected light is not so; and in the case of the liquid produced by the snow-caps of Mars, which the whole conditions of the planet show must be shallow, and also be more or less turbid, it cannot possibly be the cause of the 'deep blue' tint said to result from the melting of the snow.

But there is a very weighty argument depending on the molecular theory of gases against the polar caps of Mars being composed of frozen water at all.

The mass and elastic force of the several gases is due to the greater or less rapidity of the vibratory motion of their molecules under identical conditions. The speed of these molecular motions has been ascertained for all the chief gases, and it is found to be so great as in certain cases to enable them to overcome the force of gravity and escape from a planet's surface into space. Dr. G. Johnstone Stoney has specially investigated this subject, and he finds that the force of gravity on the earth is sufficient to retain all the gases composing its atmosphere, but not sufficient to retain hydrogen; and as a consequence, although this gas is produced in small quantities by volcanoes and by decomposing vegetation, yet no trace of it is found in our atmosphere. The moon however, having only one-eightieth the mass of the earth, cannot retain any gas, hence its airless and waterless condition.

Water Vapour cannot exist on Mars.

Now, Dr. Stoney finds that in order to retain water vapour permanently a planet must have a mass at least a quarter that of the earth. But the mass of Mars is only one-ninth that of the earth; therefore, unless there are some special conditions that prevent its loss, this gas cannot be present in the atmosphere. Mr. Lowell does not refer to this argument against his view, neither does he claim the evidence of spectroscopy in his favour. This was alleged more than thirty years ago to show the existence of water-vapour in the atmosphere of Mars, but of late years it has been doubted, and Mr. Lowell's complete silence on the subject, while laying stress on such a very weak and inconclusive argument as that from the tinge of colour that is observed a little distance from the edge of the diminishing snow-caps, shows that he himself does not think the fact to be thus proved. It he did he would hardly adduce such an argument for its presence as the following: "The melting of the caps on the one hand and their re-forming on the other affirm the presence of water-vapour in the Martian atmosphere, of whatever else that air consists" (p. 162). Yet absolutely the only proof he gives that the caps are frozen water is the almost frivolous colour-argument above referred to!

No Spectroscopic Evidence of Water Vapour.

As Sir William Huggins is the chief authority quoted for this fact, and is referred to as being almost conclusive in the third edition of Miss Clerke's History of Astronomy in 1893, I have ascertained that his opinion at the present time is that "there is no conclusive proof of the presence of aqueous vapour in the atmosphere of Mars, and that observations at the Lick Observatory (in 1895), where the conditions and instruments are of the highest order, were negative." He also informs me that Marchand at the Pic du Midi Observatory was unable to obtain lines of aqueous vapour in the spectrum of Mars; and that in 1905, Slipher, at Mr. Lowell's observatory, was unable to detect any indications of aqueous vapour in the spectrum of Mars.

It thus appears that spectroscopic observations are quite accordant with the calculations founded on the molecular theory of gases as to the absence of aqueous vapour, and therefore presumably of liquid water, from Mars. It is true that the spectroscopic argument is purely negative, and this may be due to the extreme delicacy of the observations required; but that dependent on the inability of the force of gravity to retain it is positive scientific evidence against its presence, and, till shown to be erroneous, must be held to be conclusive.

This absence of water is of itself conclusive against the existence of animal life, unless we enter the regions of pure conjecture as to the possibility of some other liquid being able to take its place, and that liquid being as omnipresent there as water is here. Mr. Lowell however never takes this ground, but bases his whole theory on the fundamental identity of the substance of the bodies of living organisms wherever they may exist in the solar system. In the next two chapters I shall discuss an equally essential condition, that of temperature, which affords a still more conclusive and even crushing argument against the suitability of Mars for the existence of organic life.

CHAPTER V.

THE TEMPERATURE OF MARS--MR. LOWELL'S ESTIMATE.

We have now to consider a still more important and fundamental question, and one which Mr. Lowell does not grapple with in this volume, the actual temperatures on Mars due to its distance from the sun and the atmospheric conditions on which he himself lays so much stress. If I am not greatly mistaken we shall arrive at conclusions on this subject which are absolutely fatal to the conception of any high form of organic life being possible on this planet.

The Problem of Terrestrial Temperatures.

In order that the problem may be understood and its importance appreciated, it is necessary to explain the now generally accepted principles as to the causes which determine the temperatures on our earth, and, presumably, on all other planets whose conditions are not wholly unlike ours. The fact of the internal heat of the earth which becomes very perceptible even at the moderate depths reached in mines and deep borings, and in the deepest mines becomes a positive inconvenience, leads many people to suppose that the surface- temperatures of the earth are partly due to this cause. But it is now generally admitted that this is not the case, the reason being that all rocks and soils, in their natural compacted state, are exceedingly bad conductors of heat.

A striking illustration of this is the fact, that a stream of lava often continues to be red hot at a few feet depth for years after the surface is consolidated, and is hardly any warmer than that of the surrounding land. A still more remarkable case is that of a glacier on the south-east side of the highest cone of Etna underneath a lava stream with an intervening bed of volcanic sand only ten feet thick. This was visited by Sir Charles Lyell in 1828, and a second time thirty years later, when he made a very careful examination of the strata, and was quite satisfied that the sand and the lava stream together had actually preserved this mass of ice, which neither the heat of the lava above it at its first outflow, nor the continued heat rising from the great volcano below it, had been able to melt or perceptibly to diminish in thirty years. Another fact that points in the same direction is the existence over the whole floor of the deepest oceans of ice-cold water, which, originating in the polar seas, owing to its greater density sinks and creeps slowly along the ocean bottom to the depths of the Atlantic and Pacific, and is not perceptibly warmed by the internal heat of the earth.

Now the solid crust of the earth is estimated as at least about twenty miles in average thickness; and, keeping in mind the other facts just referred to, we can understand that the heat from the interior passes through it with such extreme slowness as not to be detected at the surface, the varying temperatures of which are due entirely to solar heat. A large portion of this heat is stored up in the surface soil, and especially in the surface layer of the oceans and seas, thus partly equalising the temperatures of day and night, of winter and summer, so as greatly to ameliorate the rapid changes of temperature that would otherwise occur. Our dense atmosphere is also of immense advantage to us as an equaliser of temperature, charged as it almost always is with a large quantity of water-vapour. This latter gas, when not condensed into cloud, allows the solar heat to pass freely to the earth; but it has the singular and highly beneficial property of absorbing and retaining the dark or lower-grade heat given off by the earth which would otherwise radiate into space much more rapidly. We must therefore always remember that, very nearly if not quite, the whole of the warmth we experience on the earth is derived from the sun.[8]

[Footnote 8: Professor J.H. Poynting, in his lecture to the British Association at Cambridge in 1904, says: "The surface of the earth receives, we know, an amount of heat from the inside almost infinitesimal compared with that which it receives from the sun, and on the sun, therefore, we depend for our temperature."]

In order to understand the immense significance of this conclusion we must know what is meant by the whole heat or warmth; as unless we know this we cannot define what half or any other proportion of sun-heat really means.

Now I feel pretty sure that nine out of ten of the average educated public would answer the following question incorrectly: The mean temperature of the southern half of England is about 48?F. Supposing the earth received only half the sun-heat it now receives, what would then be the probable mean temperature of the South of England? The majority would, I think, answer at once--About 24?F. Nearly as many would perhaps say--48?F. is 16?above the freezing point; therefore half the heat received would bring us down to 8?above the freezing point, or 40?F. Very few, I think, would realise that our share of half the amount of sun-heat received by the earth would probably result in reducing our mean temperature to about 100?F. below the freezing point, and perhaps even lower. This is about the very lowest temperature yet experienced on the earth's surface. To understand how such results are obtained a few words must be said about the absolute zero of temperature.

The Zero of Temperature.

Heat is now believed to be entirely due to ether-vibration, which produces a correspondingly rapid vibration of the molecules of matter, causing it to expand and producing all the phenomena we term 'heat.' We can conceive this vibration to increase indefinitely, and thus there would appear to be no necessary limit to the amount of heat possible, but we cannot conceive it to decrease indefinitely at the same uniform rate, as it must soon inevitably come to nothing. Now it has been found by experiment that gases under uniform pressure expand 1/273 of their volume for each degree Centigrade of increased temperature, so that in passing from 0?C. to 273?C. they are doubled in volume. They also decrease in volume at the same rate for each degree below 0?C. (the freezing point of water). Hence if this goes on to-273?C. a gas will have no volume, or it will undergo some change of nature. Hence this is called the zero of temperature, or the temperature to which any matter falls which receives no heat from any other matter. It is also sometimes called the temperature of space, or of the ether in a state of rest, if that is possible. All the gases have now been proved to become, first liquid and then (most of them) solid, at temperatures considerably above this zero.

The only way to compare the proportional temperatures of bodies, whether on the earth or in space, is therefore by means of a scale beginning at this natural zero, instead of those scales founded on the artificial zero of the freezing point of water, or, as in Fahrenheit's, 32?below it. Only by using the natural zero and measuring continuously from it can we estimate temperatures in relative proportion to the amount of heat received. This is termed the absolute zero, and so that we start reckoning from that point it does not matter whether the scale adopted is the Centigrade or that of Fahrenheit.

The Complex Problem of Planetary Temperatures.

Now if, as is the case with Mars, a planet receives only half the amount of solar heat that we receive, owing to its greater distance from the sun, and if the mean temperature of our earth is 60?F., this is equal to 551?F. on the absolute scale. It would therefore appear very simple to halve this amount and obtain 275.5?F. as the mean temperature of that planet. But this result is erroneous, because the actual amount of sun heat intercepted by a planet is only one condition out of many that determine its resulting temperature. Radiation, that is loss of heat, is going on concurrently with gain, and the rate of loss varies with the temperature according to a law recently discovered, the loss being much greater at high temperatures in proportion to the 4th power of the absolute temperature. Then, again, the whole heat intercepted by a planet does not reach its surface unless it has no atmosphere. When it has one, much is reflected or absorbed according to complex laws dependent on the density and composition of the atmosphere. Then, again, the heat that reaches the actual surface is partly reflected and partly absorbed, according to the nature of that surface--land or water, desert or forest or snow-clad-- that part which is absorbed being the chief agent in raising the temperature of the surface and of the air in contact with it. Very important too is the loss of heat by radiation from these various heated surfaces at different rates; while the atmosphere itself sends back to the surface an ever varying portion of both this radiant and reflected heat according to distinct laws. Further difficulties arise from the fact that much of the sun's heat consists of dark or invisible rays, and it cannot therefore be measured by the quantity of light only.

From this rough statement it will be seen that the problem is an exceedingly complex one, not to be decided off-hand, or by any simple method. It has in fact been usually considered as (strictly speaking) insoluble, and only to be estimated by a more or less rough approximation, or by the method of general analogy from certain known facts. It will be seen, from what has been said in previous chapters, that Mr. Lowell, in his book, has used the latter method, and, by taking the presence of water and water-vapour in Mars as proved by the behaviour of the snow-caps and the bluish colour that results from their melting, has deduced a temperature above the freezing point of water, as prevalent in the equatorial regions permanently, and in the temperate and arctic zones during a portion of each year.

Mr. Lowell's Mathematical Investigation of the Problem.

But as this result has been held to be both improbable in itself and founded on no valid evidence, he has now, in the London, Edinburgh, and Dublin Philosophical Magazine of July 1907, published an elaborate paper of 15 pages, entitled A General Method for Evaluating the Surface-Temperatures of the Planets; with special reference to the Temperature of Mars, by Professor Percival Lowell; and in this paper, by what purports to be strict mathematical reasoning based on the most recent discoveries as to the laws of heat, as well as on measurements or estimates of the various elements and constants used in the calculations, he arrives at a conclusion strikingly accordant with that put forward in the recently published volume. Having myself neither mathematical nor physical knowledge sufficient to enable me to criticise this elaborate paper, except on a few points, I will here limit myself to giving a short account of it, so as to explain its method of procedure; after which I may add a few notes on what seem to me doubtful points; while I also hope to be able to give the opinions of some more competent critics than myself.

Mr. Lowell's Mode of Estimating the Surface-temperature of Mars.

The author first states, that Professor Young, in his General Astronomy (1898), makes the mean temperature of Mars 223.6?absolute, by using Newton's law of heat being radiated in proportion to temperature, and 363?abs. (=-96?F.) by Dulong and Petit's law; but adds, that a closer determination has been made by Professor Moulton, using Stefan's law, that radiation is as the /4th power of the temperature, whence results a mean temperature of-31?F. These estimates assume identity of atmospheric conditions of Mars and the Earth.

But as none of these estimates take account of the many complex factors which interfere with such direct and simple calculations, Mr. Lowell then proceeds to enunciate them, and work out mathematically the effects they produce:

(1) The whole radiant energy of the sun on striking a planet becomes divided as follows: Part is reflected back into space, part absorbed by the atmosphere, part transmitted to the surface of the planet. This surface again reflects a portion and only the balance left goes to warm the planet.

(2) To solve this complex problem we are helped by the albedoes or intrinsic brilliancy of the planets, which depend on the proportion of the visible rays which are reflected and which determines the comparative brightness of their respective surfaces. We also have to find the ratio of the invisible to the visible rays and the heating power of each.

(3) He then refers to the actinometer and pyroheliometer, instruments for measuring the actual heat derived from the sun, and also to the Bolometer, an instrument invented by Professor Langley for measuring the invisible heat rays, which he has proved to extend to more than three times the length of the whole heat-spectrum as previously known, and has also shown that the invisible rays contribute 68 per cent, of the sun's total energy.[9]

[Footnote 9: For a short account of this remarkable instrument, see my

Wonderful Century, new ed., pp. 143-145.]

(4) Then follows an elaborate estimate of the loss of heat during the passage of the sun's rays through our atmosphere from experiments made at different altitudes and from the estimated reflective power of the various parts of the earth's surface--rocks and soil, ocean, forest and snow--the final result being that three-fourths of the whole sun-heat is reflected back into space, forming our albedo, while only one-fourth is absorbed by the soil and goes to warm the air and determine our mean temperature.

(5) We now have another elaborate estimate of the comparative amounts of heat actually received by Mars and the Earth, dependent on their very different amounts of atmosphere, and this estimate depends almost wholly on the comparative albedoes, that of Mars, as observed by astronomers being 0.27, while ours has been estimated in a totally different way as being 0.75, whence he concludes that nearly three-fourths of the sun-heat that Mars receives reaches the surface and determines its temperature, while we get only one-fourth of our total amount. Then by applying Stefan's law, that the radiation is as the 4th power of the surface temperature, he reaches the final result that the actual heating power at the surface of Mars is considerably more than on the Earth, and would produce a mean temperature of 72?F., if it were not for the greater conservative or blanketing power of our denser and more water-laden atmosphere. The difference produced by this latter fact he minimises by dwelling on the probability of a greater proportion of carbonic-acid gas and water-vapour in the Martian atmosphere, and thus brings down the mean temperature of Mars to 48?F , which is almost exactly the same as that of the southern half of England. He has also, as the result of observations, reduced the probable density of the atmosphere of Mars to 2-1/2 inches of mercury, or only one-twelfth of that of the Earth.

Critical Remarks on Mr. Lowell's Paper.

The last part of this paper, indicated under pars. 4 and 5, is the most elaborate, occupying eight pages, and it contains much that seems uncertain, if not erroneous. In particular, it seems very unlikely that under a clear sky over the whole earth we should only receive at the sea-level 0.23 of the solar rays which the earth intercepts (p. 167). These data largely depend on observations made in California and other parts of the southern United States, where the lower atmosphere is exceptionally dust-laden. Till we have similar observations made in the tropical forest-regions, which cover so large an area, and where the atmosphere is purified by frequent rains, and also on the prairies and the great oceans, we cannot trust these very local observations for general conclusions affecting the whole earth. Later, in the same article (p. 170), Mr. Lowell says: "Clouds transmit approximately 20 per cent. of the heat reaching them: a clear sky at sea-level 60 per cent. As the sky is half the time cloudy the mean transmission is 35 per cent." These statements seem incompatible with that quoted above.

The figure he uses in his calculations for the actual albedo of the earth, 0.75, is also not only improbable, but almost self-contradictory, because the albedo of cloud is 0.72, and that of the great cloud-covered planet, Jupiter, is given by Lowell as 0.75, while Zollner made it only 0.62. Again, Lowell gives Venus an albedo of 0.92, while Zollner made it only 0.50 and Mr. Gore 0.65. This shows the extreme uncertainty of these estimates, while the fact that both Venus and Jupiter are wholly cloud-covered, while we are only half-covered, renders it almost certain that our albedo is far less than Mr. Lowell makes it. It is evident that mathematical calculations founded upon such uncertain data cannot yield trustworthy results. But this is by no means the only case in which the data employed in this paper are of uncertain value. Everywhere we meet with figures of somewhat doubtful accuracy. Here we have somebody's 'estimate' quoted, there another person's 'observation,' and these are adopted without further remark and used in the various calculations leading to the result above quoted. It requires a practised mathematician, and one fully acquainted with the extensive literature of this subject, to examine these various data, and track them through the maze of formulae and figures so as to determine to what extent they affect the final result.

There is however one curious oversight which I must refer to, as it is a point to which I have given much attention. Not only does Mr. Lowell assume, as in his book, that the 'snows' of Mars consist of frozen water, and that therefore there is water on its surface and water-vapour in its atmosphere, not only does he ignore altogether Dr. Johnstone Stoney's calculations with regard to it, which I have already referred to, but he uses terms that imply that water-vapour is one of the heavier components of our atmosphere. The passage is at p. 168 of the Philosophical Magazine. After stating that, owing to the very small barometric pressure in Mars, water would boil at 110?F., he adds: "The sublimation at lower temperatures would be correspondingly increased. Consequently the amount of water-vapour in the Martian air must on that score be relatively greater than our own." Then follows this remarkable passage: "Carbon-dioxide, because of its greater specific gravity, would also be in relatively greater amount so far as this cause is considered. For the planet would part, caeteris paribus, with its lighter gases the quickest. Whence as regards both water-vapour and carbon-dioxide we have reason to think them in relatively greater quantity than in our own air at corresponding barometric pressure." I cannot understand this passage except as implying that 'water-vapour and carbon-dioxide' are among the heavier and not among the lighter gases of the atmosphere--those which the planet 'parts with quickest.' But this is just what water-vapour is, being a little less than two-thirds the weight of air (0.6225), and one of those which the planet would part with the quickest, and which, according to Dr. Johnstone Stoney, it loses altogether. * * * * *

Note on Professor Lowell's article in the Philosophical Magazine; by J.H. Poynting, F.R.S., Professor of Physics in the University of Birmingham.

"I think Professor Lowell's results are erroneous through his neglect of the heat stored in the air by its absorption of radiation both from the sun and from the surface. The air thus heated radiates to the surface and keeps up the temperature. I have sent to the Philosophical Magazine a paper in which I think it is shown that when the radiation by the atmosphere is taken into account the results are entirely changed. The temperature of Mars, with

Professor Lowell's data, still comes out far below the freezing-point--still further below than the increased distance alone would make it. Indeed, the lower temperature on elevated regions of the earth's surface would lead us to expect this. I think it is impossible to raise the temperature of Mars to anything like the value obtained by Professor Lowell, unless we assume some quality in his atmosphere entirely different from any found in our own atmosphere." J.H. POYNTING. October 19, 1907.

CHAPTER VI.

A NEW ESTIMATE OF THE TEMPERATURE OF MARS.

When we are presented with a complex problem depending on a great number of imperfectly ascertained data, we may often check the results thus obtained by the comparison of cases in which some of the more important of these data are identical, while others are at a maximum or a minimum. In the present case we can do this by a consideration of the Moon as compared with the Earth and with Mars.

Langley's Determination of the Moon's Temperature.

In the moon we see the conditions that prevail in Mars both exaggerated and simplified. Mars has a very scanty atmosphere, the moon none at all, or if there is one it is so excessively scanty that the most refined observations have not detected it. All the complications arising from the possible nature of the atmosphere, and its complex effects upon reflection, absorption, and radiation are thus eliminated. The mean distance of the moon from the sun being identical with that of the earth, the total amount of heat intercepted must also be identical; only in this case the whole of it reaches the surface instead of one-fourth only, according to Mr. Lowell's estimate for the earth.

Now, by the most refined observations with his Bolometer, Mr. Langley was able to determine the temperature of the moon's surface exposed to undimmed sunshine for fourteen days together; and he found that, even in that portion of it on which the sun was shining almost vertically, the temperature rarely rose above the freezing point of water. However extraordinary this result may seem, it is really a striking confirmation of the accuracy of the general laws determining temperature which I have endeavoured to explain in the preceding chapter. For the same surface which has had fourteen days of sunshine has also had a preceding fourteen days of darkness, during which the heat which it had accumulated in its surface layers would have been lost by free radiation into stellar space. It thus acquires during its day a maximum temperature of only 491?F. absolute, while its minimum, after 14 days' continuous radiation, must be very low, and is, with much reason, supposed to approach the absolute zero.

Rapid Loss of Heat by Radiation on the Earth.

In order better to comprehend what this minimum may be under extreme conditions, it will be useful to take note of the effects it actually produces on the earth in places where the conditions are nearest to those existing on the moon or on Mars, though never quite equalling, or even approaching very near them. It is in our great desert regions, and especially on high plateaux, that extreme aridity prevails, and it is in such districts that the differences between day and night temperatures reach their maximum. It is stated by geographers that in parts of the Great Sahara the surface temperature is sometimes 150?F., while during the night it falls nearly or quite to the freezing point--a difference of 118 degrees in little more than 12 hours.[10] In the high desert plains of Central Asia the extremes are said to be even greater.[11] Again, in his Universal Geography, Reclus states that in the Armenian Highlands the thermometer oscillates between 13?F. and 112 癨. We may therefore, without any fear of exaggeration, take it as proved that a fall of 100?F. in twelve or fifteen hours not infrequently occurs where there is a very dry and clear atmosphere permitting continuous insolation by day and rapid radiation by night.

[Footnote 10: Keith Johnston's 'Africa' in Stanford's Compendium.]

Now, as it is admitted that our dense atmosphere, however dry and clear, absorbs and reflects some considerable portion of the solar heat, we shall certainly underestimate the radiation from the moon's surface during its long night if we take as the basis of our calculation a lowering of temperature amounting to 100?F. during twelve hours, as not unfrequently occurs with us. Using these data--with Stefan's law of decrease of radiation as the 4th power of the temperature--a mathematical friend finds that the temperature of the moon's surface would be reduced during the lunar night to nearly 200?F. absolute (equal to-258?F.).

More Rapid Loss of Heat by the Moon.

Although such a calculation as the above may afford us a good approximation to the rate of loss of heat by Mars with its very scanty atmosphere, we have now good evidence that in the case of the moon the loss is much more rapid. Two independent workers have investigated this subject with very accordant results--Dr. Boeddicker, with Lord Rosse's 3-foot reflector and a Thermopile to measure the heat, and Mr. Frank Very, with a glass reflector of 12 inches diameter and the Bolometer invented by Mr. Langley. The very striking and unexpected fact in which these observers agree is the sudden disappearance of much of the stored-up heat during the comparatively short duration of a total eclipse of the moon--less than two hours of complete darkness, and about twice that period of partial obscuration.

Dr. Boeddicker was unable to detect any appreciable heat at the period of greatest obscuration; but, owing to the extreme sensitiveness of the Bolometer, Mr. Very ascertained that those parts of the surface which had been longest in the shadow still emitted heat "to the amount of one per cent. of the heat to be expected from the full moon." This however is the amount of radiation measured by the Bolometer, and to get the temperature of the radiating surface we must apply Stefan's law of the 4th power. Hence the temperature of the moon's dark surface will be the [fourth root of (1 over 100)] = 1 over 3.2 [A] of the highest temperature (which we may take at the freezing-point, 491?F. abs.), or 154?F. abs., just below the liquefaction point of air. This is about 50?lower than the amount found by calculation from our most rapid radiation; and as this amount is produced in a few hours, it is not too much to expect that, when continued for more than two weeks (the lunar night), it might reach a temperature sufficient to liquefy hydrogen (60?F. abs.), or perhaps even below it.

[Note A: LaTex markup $\root 4 \of {1 \over 100} = {1 \over 3.2}$]

Theory of the Moon's Origin.

This extremely rapid loss of heat by radiation, at first sight so improbable as to be almost incredible, may perhaps be to some extent explained by the physical constitution of the moon's surface, which, from a theoretical point of view, does not appear to have received the attention it deserves. It is clear that our satellite has been long subjected to volcanic eruptions over its whole visible face, and these have evidently been of an explosive nature, so as to build up the very lofty cones and craters, as well as thousands of smaller ones, which, owing to the absence of any degrading or denuding agencies, have remained piled up as they were first formed.

This highly volcanic structure can, I think, be well explained by an origin such as that attributed to it by Sir George Darwin, and which has been so well described by Sir Robert Ball in his small volume, Time and Tide. These astronomers adduce strong evidence that the earth once rotated so rapidly that the equatorial protuberance was almost at the point of separation from the planet as a ring. Before this occurred, however, the tension was so great that one large portion of the protuberance where it was weakest broke away, and began to move around the earth at some considerable distance from it. As about 1/50 of the bulk of the earth thus escaped, it must have consisted of a considerable portion of the solid crust and a much larger quantity of the liquid or semi-liquid interior, together with a proportionate amount of the gases which we know formed, and still form, an important part of the earth's substance.

As the surface layers of the earth must have been the lightest, they would necessarily, when broken up by this gigantic convulsion, have come together to form the exterior of the new satellite, and be soon adjusted by the forces of gravity and tidal disturbance into a more or less irregular spheroidal form, all whose interstices and cavities would be filled up and connected together by the liquid or semi-liquid mass forced up between them. Thence-forward, as the moon increased its distance and reduced its time of rotation, in the way explained by Sir Robert Ball, there would necessarily commence a process of escape of the imprisoned gases at every fissure and at all points and lines of weakness, giving rise to numerous volcanic outlets, which, being subjected only to the small force of lunar gravity (only one-sixth that of the earth), would, in the course of ages, pile up those gigantic cones and ridges which form its great characteristic.

But this small gravitative power of the moon would prevent its retaining on its surface any of the gases forming our atmosphere, which would all escape from it and probably be recaptured by the earth. By no process of external aggregation of solid matter to such a relatively small amount as that forming the moon, even if the aggregation was so violent as to produce heat enough to cause liquefaction, could any such long-continued volcanic action arise by gradual cooling, in the absence of internal gases. There might be fissures, and even some outflows of molten rock; but without imprisoned gases, and especially without water and water-vapour producing explosive outbursts, could any such amount of scoriae and ashes be produced as were necessary for the building up of the vast volcanic cones, craters, and craterlets we see upon the moon's surface.

I am not aware that either Sir Robert Ball or Sir George Darwin have adduced this highly volcanic condition of the moon's surface as a phenomenon which can only be explained by our satellite having been thrown off a very much larger body, whose gravitative force was sufficient to acquire and retain the enormous quantity of gases and of water which we possess, and which are absolutely essential for that special form of cone-building volcanic action which the moon exhibits in so pre-eminent a degree. Yet it seems to me clear, that some such hypothetical origin for our satellite would have had to be assumed if Sir George Darwin had not deduced it by means of purely mathematical argument based upon astronomical facts.

Returning now to the problem of the moon's temperature, I think the phenomena this presents may be in part due to the mode of formation here described. For, its entire surface being the result of long-continued gaseous explosions, all the volcanic products--scoriae, pumice, and ashes--would necessarily be highly porous throughout; and, never having been compacted by water-action, as on the earth, and there having been no winds to carry the finer dust so as to fill up their pores and fissures, the whole of the surface material to a very considerable depth must be loose and porous to a high degree. This condition has been further increased owing to the small power of gravity and the extreme irregularity of the surface, consisting very largely of lofty cones and ridges very loosely piled up to enormous heights.

Now this condition of the substance of the moon's surface is such as would produce a high specific heat, so that it would absorb a large amount of heat in proportion to the rise of temperature produced, the heat being conducted downwards to a considerable depth. Owing, however, to the total absence of atmosphere radiation would very rapidly cool the surface, but afterwards more slowly, both on account of the action of Stefan's law and because the heat stored up in the deeper portions could be carried to the surface by conduction only, and with extreme slowness.

Very's Researches on the Moon's Heat.

The results of the eclipse observations are supported by the detailed examination of the surface-temperature of the moon by Mr. Very in his Prize Essay on the Distribution of the Moon's Heat (published by the Utrecht Society of Arts and Sciences in 1891). He shows, by a diagram of the 'Phasecurve,' that at the commencement of the Lunar day the surface just within the illuminated limb has acquired about 1/7 of its maximum temperature, or about 70?F. abs. As the surface exposed to the Bolometer at each observation is about 1/30 of the moon's surface, and in order to ensure accuracy the instrument has to be directed to a spot lying wholly within the edge of the moon, it is evident that the surface measured has already been for several hours exposed to oblique sunshine. The curve of temperature then rises gradually and afterwards more rapidly, till it attains its maximum (of about +30 to 40?F.) a few hours before noon. This, Mr. Very thinks, is due to the fact that the half of the moon's face first illuminated for us has, on the average, a darker surface than that of the afternoon, or second quarter, during which the curve descends not quite so rapidly, the temperature near sunset being only a little higher than that near sunrise. This rapid fall while exposed to oblique sunshine is quite in harmony with the rapid loss of heat during the few hours of darkness during an eclipse, both showing the prepotency of radiation over insolation on the moon.

Two other diagrams show the distribution of heat at the time of full-moon, one half of the curve showing the temperatures along the equator from the edge of the disc to the centre, the other along a meridian from this centre to the pole. This diagram (here reproduced) exhibits the quick rise of temperature of the oblique rim of the moon and the nearly uniform heat of the central half of its surface; the diminution of heat towards the pole, however, is slower for the first half and more rapid for the latter portion.

It is an interesting fact that the temperature near the margin of the full-moon increases towards the centre more rapidly than it does when the same parts are observed during the early phases of the first quarter. Mr. Very explains this difference as being due to the fact that the full-moon to its very edges is fully illuminated, all the shadows of the ridges and mountains being thrown vertically or obliquely behind them. We thus measure the heat reflected from the whole visible surface. But at new moon, and somewhat beyond the first quarter, the deep shadows thrown by the smallest cones and ridges, as well as by the loftiest mountains, cover a considerable portion of the visible surface, thus largely reducing the quantity of light and heat reflected or radiated in our direction. It is only at the full, therefore, that the maximum temperature of the whole lunar surface can be measured. It must be considered a proof of the delicacy of the heat-measuring instruments that this difference in the curves of temperature of the different parts of the moon's surface and under different conditions is so clearly shown.

The Application of the Preceding Results to the Case of Mars.

This somewhat lengthy account of the actual state of the moon's surface and temperature is of very great importance in our present enquiry, because it shows us the extraordinary difference in mean and extreme temperatures of two bodies situated at the same distance from the sun, and therefore receiving exactly the same amount of solar heat per unit of surface. We have learned also what are the main causes of this almost incredible difference, namely: (1) a remarkably rugged surface with porous and probably cavernous rock-texture, leading to extremely rapid radiation of heat in the one; as compared with a comparatively even and well-compacted surface largely clad with vegetation, leading to comparatively slow and gradual loss by radiation in the other: and (2), these results being greatly intensified by the total absence of a protecting atmosphere in the former, while a dense and cloudy atmosphere with an ever-present supply of water-vapour, accumulates and equalises the heat received by the latter.

The only other essential difference in the two bodies which may possibly aid in the production of this marvellous result, is the fact of our day and night having a mean length of 12 hours, while those of the moon are about 14-1/2 of our days. But the altogether unexpected fact, in which two independent enquirers agree, that during the few hours' duration of a total eclipse of the moon so large a proportion of the heat is lost by radiation renders it almost certain that the resulting low temperature would be not very much less if the moon had a day and night the same length as our own.

The great lesson we learn by this extreme contrast of conditions supplied to us by nature, as if to enable us to solve some of her problems, is, the overwhelming importance, first, of a dense and well-compacted surface, due to water-action and strong gravitative force; secondly, of a more or less general coat of vegetation; and, thirdly, of a dense vapour-laden atmosphere. These three favourable conditions result in a mean temperature of about +60?F. with a range seldom exceeding 40?above or below it, while over more than half the land-surface of the earth the temperature rarely falls below the freezing point. On the other hand, we have a globe of the same materials and at the same distance from the sun, with a maximum temperature of freezing water, and a minimum not very far from the absolute zero, the monthly mean being probably much below the freezing point of carbonic-acid gas--a difference entirely due to the absence of these three favourable conditions.

The Special Features of Mars as influencing Temperature.

Coming now to the special feature of Mars and its probable temperature, we find that most writers have arrived at a very different conclusion from that of Mr. Lowell, who himself quotes Mr. Moulton as an authority who 'recently, by the application of Stefan's law,' has found the mean temperature of this planet to be-35?F. Again, Professor J.H. Poynting, in his lecture on 'Radiation in the Solar System,' delivered before the British Association at Cambridge in 1904, gave an estimate of the mean temperature of the planets, arrived at from measurements of the sun's emissive power and the application of Stefan's law to the distances of the several planets, and he thus finds the earth to have a mean temperature of 17?C. (=62-1/2?F.) and Mars one of-38?C. (=-36-1/2?F.), a wonderfully close approximation to the mean temperature of the earth as determined by direct measurement, and therefore, presumably, an equally near approximation to that of Mars as dependent on distance from the sun, and 'on the supposition that it is earth-like in all its conditions.'

But we know that it is far from being earth-like in the very conditions which we have found to be those which determine the extremely different temperatures of the earth, and moon; and, as regards each of these, we shall find that, so far as it differs from the earth, it approximates to the less favourable conditions that prevail in the moon. The first of these conditions which we have found to be essential in regulating the absorption and radiation of heat, and thus raising the mean temperature of a planet, is a compact surface well covered with vegetation, two conditions arising from, and absolutely dependent on, an ample amount of water. But Mr. Lowell himself assures us, as a fact of which he has no doubt, that there are no permanent bodies of water, great or small, upon Mars; that rain, and consequently rivers, are totally wanting; that its sky is almost constantly clear, and that what appear to be clouds are not formed of water-vapour but of dust. He dwells, emphatically, on the terrible desert conditions of the greater part of the surface of the planet.

That being the case now, we have no right to assume that it has ever been otherwise; and, taking full account of the fact, neither denied nor disputed by Mr. Lowell, that the force of gravity on Mars is not sufficient to retain water-vapour in its atmosphere, we must conclude that the surface of that planet, like that of the moon, has been moulded by some form of volcanic action modified probably by wind, but not by water. Adding to this, that the force of gravity on Mars is nearer that of the moon than to that of the earth, and we may r reasonably conclude that its surface is formed of volcanic matter in a light and porous condition, and therefore highly favourable for the rapid loss of surface heat by radiation. The surface-conditions of Mars are therefore, presumably, much more like those of the moon than like those of the earth.

The next condition favourable to the storing up of heat--a covering of vegetation--is almost certainly absent from Mars except, possibly, over limited areas and for short periods. In this feature also the surface of Mars approximates much nearer to lunar than to earth-conditions. The third condition--a dense, vapour-laden atmosphere--is also wanting in Mars. For although it possesses an atmosphere it is estimated by Mr. Lowell (in his latest article) to have a pressure equivalent to only 2 1/2 inches of mercury with us, giving it a density of only one-twelfth part that of ours; while aqueous vapour, the chief accumulator of heat, cannot permanently exist in it, and, notwithstanding repeated spectroscopic observations for the purpose of detecting it, has never been proved to exist.

I submit that I have now shown from the statements--and largely as the result of the long-continued observations--of Mr. Lowell himself, that, so far as the physical conditions of Mars are known to differ from those of the earth, the differences are all unfavourable to the conservation and favourable to the dissipation of the scanty heat it receives from the sun--that they point unmistakeably towards the temperature conditions of the moon rather than to those of the earth, and that the cumulative effect of these adverse conditions, acting upon a heat-supply, reduced by solar distance to less than one-half of ours, must result in a mean temperature (as well as in the extremes) nearer to that of our satellite than to that of our own earth.

Further Criticism of Mr. Lowell's Article.

We are now in a position to test some further conclusions of Mr. Lowell's Phil. Mag. article by comparison with actual phenomena. We have seen, in the outline I have given of this article, that he endeavours to show how the small amount of solar heat received by Mars is counterbalanced, largely by the greater transparency to light and heat of its thin and cloudless atmosphere, and partially also by a greater conservative or 'blanketing' power of its atmosphere due to the presence in it of a large proportion of carbonic acid gas and aqueous vapour. The first of these statements may be admitted as a fact which he is entitled to dwell upon, but the second--the presence of large quantities of carbon-dioxide and aqueous vapour is a pure hypothesis unsupported by any item of scientific evidence, while in the case of aqueous vapour it is directly opposed to admitted results founded upon the molecular theory of gaseous elasticity. But, although Mr. Lowell refers to the conservative or 'blanketing' effect of the earth's atmosphere, he does not consider or allow for its very great cumulative effect, as is strikingly shown by the comparison with the actual temperature conditions of the moon. This cumulative effect is due to the continuous reflection and radiation of heat from the clouds as well as from the vapour-laden strata of air in our lower atmosphere, which latter, though very transparent to the luminous and accompanying heat rays of the sun, are opaque to the dark heat-rays whether radiated or reflected from the earth's surface. We are therefore in a position strictly comparable with that of the interior of some huge glass house, which not only becomes intensely heated by the direct rays of the sun, but also to a less degree by reflected rays from the sky and those radiated from the clouds, so that even on a cloudy or misty day its temperature rises many degrees above that of the outer air. Such a building, if of large size, of suitable form, and well protected at night by blinds or other covering, might be so arranged as to accumulate heat in its soil and walls so as to maintain a tolerably uniform temperature though exposed to a considerable range of external heat and cold. It is to such a power of accumulation of heat in our soil and lower atmosphere that we must impute the overwhelming contrast between our climate and that of the moon. With us, the solar heat that penetrates our vapour-laden and cloudy atmosphere is shut in by that same atmosphere, accumulates there for weeks and months together, and can only slowly escape. It is this great cumulative power which Mr. Lowell has not taken account of, while he certainly has not estimated the enormous loss of heat by free radiation, which entirely neutralises the effects of increase of sun-heat, however great, when these cumulative agencies are not present.[12]

[Footnote 12: The effects of this 'cumulative' power of a dense atmosphere are further discussed and illustrated in the last chapter of this book, where I show that the universal fact of steadily diminishing temperatures at high altitudes is due solely to the diminution of this cumulative power of our atmosphere, and that from this cause alone the temperature of Mars must be that which would be found on a lofty plateau about 18,000 feet higher than the average of the peaks of the Andes!]

Temperature on Polar Regions of Mars.

There is also a further consideration which I think Mr. Lowell has altogether omitted to discuss. Whatever may be the mean temperature of Mars, we must take account of the long nights in its polar and high-temperate latitudes, lasting nearly twice as long as ours, with the resulting lowering of temperature by radiation into a constantly clear sky. Even in Siberia, in Lat. 67-1/2. a cold of-88. has been attained; while over a large portion of N. Asia and America above 60 Lat. the mean January temperature is from-30. to-60., and the whole subsoil is permanently frozen from a depth of 6 or 7 feet to several hundreds. But the winter temperatures, over the same latitudes in Mars, must be very much lower; and it must require a proportionally larger amount of its feeble sun-heat to raise the surface even to the freezing-point, and an additional very large amount to melt any considerable depth of snow. But this identical area, from a little below 60 to the pole, is that occupied by the snow-caps of Mars, and over the whole of it the winter temperature must be far lower than the earth-minimum of-88. Then, as the Martian summer comes on, there is less than half the sun-heat available to raise this low temperature after a winter nearly double the length of ours. And when the summer does come with its scanty sun-heat, that heat is not accumulated as it is by our dense and moisture-laden atmosphere, the marvellous effects of which we have already shown. Yet with all these adverse conditions, each assisting the other to produce a climate approximating to that which the earth would have if it had no atmosphere (but retaining our superiority over Mars in receiving double the amount of sun-heat), we are asked to accept a mean temperature for the more distant planet almost exactly the same as that of mild and equable southern England, and a disappearance of the vast snowfields of its polar regions as rapid and complete as what occurs with us! If the moon, even at its equator, has not its temperature raised above the freezing-point of water, how can the more distant Mars, with its oblique noon-day sun falling upon the snow-caps, receive heat enough, first to raise their temperature to 32?F., and then to melt with marked rapidity the vast frozen plains of its polar regions?

Mr. Lowell is however so regardless of the ordinary teachings of meteorological science that he actually accounts for the supposed mild climate of the polar regions of Mars by the absence of water on its surface and in its atmosphere. He concludes his fifth chapter with the following words: "Could our earth but get rid of its oceans, we too might have temperate regions stretching to the poles." Here he runs counter to two of the best-established laws of terrestrial climatology-- the wonderful equalising effects of warm ocean-currents which are the chief agents in diminishing polar cold; the equally striking effects of warm moist winds derived from these oceans, and the great storehouse of heat we possess in our vapour-laden atmosphere, its vapour being primarily derived from these same oceans! But, in Mr. Lowell's opinion, all our meteorologists are quite mistaken. Our oceans are our great drawbacks. Only get rid of them and we should enjoy the exquisite climate of Mars--with its absence of clouds and fog, of rain or rivers, and its delightful expanses of perennial deserts, varied towards the poles by a scanty snow-fall in winter, the melting of which might, with great care, supply us with the necessary moisture to grow wheat and cabbages for about one-tenth, or more likely one-hundredth, of our present population. I hope I may be excused for not treating such an argument seriously. The various considerations now advanced, especially those which show the enormous cumulative and conservative effect of our dense and water-laden atmosphere, and the disastrous effect--judging by the actual condition of the moon--which the loss of it would have upon our temperature, seem to me quite sufficient to demonstrate important errors in the data or fallacies in the complex mathematical argument by which Mr. Lowell has attempted to uphold his views as to the temperature and consequent climatic conditions of Mars. In concluding this portion of my discussion of the problem of Mars, I wish to call attention to the fact that my argument, founded upon a comparison of the physical conditions of the earth and moon with those of Mars, is dependent upon a small number of generally admitted scientific facts; while the conclusions drawn from those facts are simple and direct, requiring no mathematical knowledge to follow them, or to appreciate their weight and cogency. I claim for them, therefore, that they are in no degree speculative, but in their data and methods exclusively scientific. In the next chapter I will put forward a suggestion as to how the very curious markings upon the surface of Mars may possibly be interpreted, so as to be in harmony with the planet's actual physical condition and its not improbable origin and past history.

CHAPTER VII.

A SUGGESTION AS TO THE 'CANALS' OF MARS.

The special characteristics of the numerous lines which intersect the whole of the equatorial and temperate regions of Mars are, their straightness combined with their enormous length. It is this which has led Mr. Lowell to term them 'non-natural features.' Schiaparelli, in his earlier drawings, showed them curved and of comparatively great width. Later, he found them to be straight fine lines when seen under the best conditions, just as Mr. Lowell has always seen them in the pure atmosphere of his observatory. Both of these observers were at first doubtful of their reality, but persistent observation continued at many successive oppositions compelled acceptance of them as actual features of the planet's disc. So many other observers have now seen them that the objection of unreality seems no longer valid.

Mr. Lowell urges, however, that their perfect straightness, their extreme tenuity, their uniformity throughout their whole length, the dual character of many of them, their relation to the 'oases' and the form and position of these round black spots, are all proofs of artificiality and are suggestive of design. And considering that some of them are actually as long as from Boston to San Francisco, and relatively to their globe as long as from London to Bombay, his objection that "no natural phenomena within our knowledge show such regularity on such a scale" seems, at first, a mighty one.

It is certainly true that we can point to nothing exactly like them either on the earth or on the moon, and these are the only two planetary bodies we are in a position to compare with Mars. Yet even these do, I think, afford us some hints towards an interpretation of the mysterious lines. But as our knowledge of the internal structure and past history even of our earth is still imperfect, that of the moon only conjectural, and that of Mars a perfect blank, it is not perhaps surprising that the surface-features of the latter do not correspond with those of either of the others.

Mr. Pickering's Suggestion.

The best clue to a natural interpretation of the strange features of the surface of Mars is that suggested by the American astronomer Mr. W.H. Pickering in Popular Astronomy (1904). Briefly it is, that both the 'canals' of Mars and the rifts as well as the luminous streaks on the moon are cracks in the volcanic crust, caused by internal stresses due to the action of the heated interior. These cracks he considers to be symmetrically arranged with regard to small 'craterlets' (Mr. Lowell's 'oases') because they have originated from them, just as the white streaks on the moon radiate from the larger craters as centres. He further supposes that water and carbon-dioxide issue from the interior into these fissures, and, in conjunction with sunlight, promote the growth of vegetation. Owing to the very rare atmosphere, the vapours, he thinks, would not ascend but would roll down the outsides of the craterlets and along the borders of the canals, thus irrigating the immediate vicinity and serving to promote the growth of some form of vegetation which renders the canals and oases visible.[13]

[Footnote 13: Nature, vol. 70, p. 536.]

This opinion is especially important because, next to Mr. Lowell, Mr. Pickering is perhaps the astronomer who has given most attention to Mars during the last fifteen years. He was for some time at Flagstaff with Mr. Lowell, and it was he who discovered the oases or craterlets, and who originated the idea that we did not see the 'canals' themselves but only the vegetable growth on their borders. He also observed Mars in the Southern Hemisphere at Arequipa; and he has since made an elaborate study of the moon by means of a specially constructed telescope of 135 feet focal length, which produced a direct image on photographic plates nearly 16 inches in diameter.[14]

[Footnote 14: Nature, vol. 70, May 5, p.xi, supplement.]

It is clear therefore that Mr. Lowell's views as to the artificial nature of the 'canals' of Mars are not accepted by an astronomer of equal knowledge and still wider experience. Yet Professor Pickering's alternative view is more a suggestion than an explanation, because there is no attempt to account for the enormous length and perfect straightness of the lines on Mars, so different from anything that is found either on our earth or on the moon. There must evidently be some great peculiarity of structure or of conditions on Mars to account for these features, and I shall now attempt to point out what this peculiarity is and how it may have arisen.

The Meteoritic Hypothesis.

During the last quarter of a century a considerable change has come over the opinions of astronomers as regards the probable origin of the Solar System. The large amount of knowledge of the stellar universe, and especially of nebulae, of comets and of meteor-streams which we now possess, together with many other phenomena, such as the constitution of Saturn's rings, the great number and extent of the minor planets, and generally of the vast amount of matter in the form of meteor-rings and meteoric dust in and around our system, have all pointed to a different origin for the planets and their satellites than that formulated by Laplace as the Nebular hypothesis.

It is now seen more clearly than at any earlier period, that most of the planets possess special characteristics which distinguish them from one another, and that such an origin as Laplace suggested--the slow cooling and contraction of one vast sun-mist or nebula, besides presenting inherent difficulties--many think them impossibilities--in itself does not afford an adequate explanation of these peculiarities. Hence has arisen what is termed the Meteoritic theory, which has been ably advocated for many years by Sir Norman Lockyer, and with some unimportant modifications is now becoming widely accepted. Briefly, this theory is, that the planets have been formed by the slow aggregation of solid particles around centres of greatest condensation; but as many of my readers may be altogether unacquainted with it, I will here give a very clear statement of what it is, from Professor J.W. Gregory's presidential address to the Geological Section of the British

Association of the present year. He began by saying that these modern views were of far more practical use to men of science than that of Laplace, and that they give us a history of the world consistent with the actual records of geology. He then continues:

"According to Sir Norman Lockyer's Meteoritic Hypothesis, nebulae, comets, and many so-called stars consist of swarms of meteorites which, though normally cold and dark, are heated by repeated collisions, and so become luminous. They may even be volatilised into glowing meteoric vapour; but in time this heat is dissipated, and the force of gravity condenses a meteoritic swarm into a single globe. 'Some of the swarms are,' says Lockyer, 'truly members of the solar system,' and some of these travel round the sun in nearly circular orbits, like planets. They may be regarded as infinitesimal planets, and so Chamberlain calls them 'planetismals.'

"The planetismal theory is a development of the meteoritic theory, and presents it in an especially attractive guise. It regards meteorites as very sparsely distributed through space, and gravity as powerless to collect them into dense groups. So it assigns the parentage of the solar system to a spiral nebula composed of planetismals, and the planets as formed from knots in the nebula, where many planetismals had been concentrated near the intersections of their orbits. These groups of meteorites, already as dense as a swarm of bees, were then packed closer by the influence of gravity, and the contracting mass was heated by the pressure, even above the normal melting-point of the material, which was kept rigid by the weight of the overlying layers."

Now, adopting this theory as the last word of science upon the subject of the origin of planets, we see that it affords immense scope for diversity in results depending on the total amount of matter available within the range of attraction of an incipient planetary mass, and the rates at which this matter becomes available. By a special combination of these two quantities (which have almost certainly been different for each planet) I think we may be able to throw some light upon the structure and physical features of Mars.

The Probable Mode of Origin of Mars.

This planet, lying between two of much greater mass, has evidently had less material from which to be formed by aggregation; and if we assume--as in the absence of evidence to the contrary we have a right to do--that its beginnings were not much later (or earlier) than those of the earth, then its smaller size shows that it has in all probability aggregated very much more slowly. But the internal heat acquired by a planet while forming in this manner will depend upon the rate at which it aggregates and the velocity with which the planetismals' fall into it, and this velocity will increase with its mass and consequent force of gravity. In the early stages of a planet's growth it will probably remain cold, the small amount of heat produced by each impact being lost by radiation before the next one occurs; and with a small and slowly aggregating planet this condition will prevail till it approaches its full size. Then only will its gravitative force be sufficient to cause incoming matter to fall upon it with so powerful an impact as to produce intense heat. Further, the compressive force of a small planet will be a less effective heat-producing agency than in the case of a larger one.

The earth we know has acquired a large amount of internal heat, probably sufficient to liquefy its whole interior; but Mars has only one-ninth part the mass of the earth, and it is quite possible, and even probable, that its comparatively small attractive force would never have liquefied or even permanently heated the more central portions of its mass. This being admitted, I suggest the following course of events as quite possible, and not even improbable, in the case of this planet. During the whole of its early growth, and till it acquired nearly its present diameter, its rate of aggregation was so slow that the planetismals falling upon it, though they might have been heated and even partially liquefied by the impact, were never in such quantity as to produce any considerable heating effect on the whole mass, and each local rise of temperature was soon lost by radiation. The planet thus grew as a solid and cold mass, compacted together by the impact of the incoming matter as well as by its slowly increasing gravitative force. But when it had attained to within perhaps 100, perhaps 50 miles, or less, of its present diameter, a great change occurred in the opportunity for further growth. Some large and dense swarm of meteorites, perhaps containing a number of bodies of the size of the asteroids, came within the range of the sun's attraction and were drawn by it into an orbit which crossed that of Mars at such a small angle that the planet was able at each revolution to capture a considerable number of them. The result might then be that, as in the case of the earth, the continuous inpour of the fresh matter first heated, and later on liquefied the greater part of it as well perhaps as a thin layer of the planet's original surface; so that when in due course the whole of the meteor-swarm had been captured, Mars had acquired its present mass, but would consist of an intensely heated, and either liquid or plastic thin outer shell resting upon a cold and solid interior.

The size and position of the two recently discovered satellites of Mars, which are believed to be not more than ten miles in diameter, the more remote revolving around its primary very little slower than the planet rotates, while the nearer one, which is considerably less distant from the planet's surface than its own antipodes and revolves around it more than three times during the Martian day, may perhaps be looked upon as the remnants of the great meteor-swarm which completed the Martian development, and which are perhaps themselves destined at some distant period to fall into the planet. Should future astronomers witness the phenomenon the effect produced upon its surface would be full of instruction.

As the result of such an origin as that suggested, Mars would possess a structure which, in the essential feature of heat-distribution, would be the very opposite of that which is believed to characterise the earth, yet it might have been produced by a very slight modification of the same process. This peculiar heat-distribution, together with a much smaller mass and gravitative force, would lead to a very different development of the surface and an altogether diverse geological history from ours, which has throughout been profoundly influenced by its heated interior, its vast supply of water, and the continuous physical and chemical reactions between the interior and the crust.

These reactions have, in our case, been of substantially the same nature, and very nearly of the same degree of intensity throughout the whole vast eons of geological time, and they have resulted in a wonderfully complex succession of rock-formations--volcanic, plutonic, and sedimentary--more or less intermingled throughout the whole series, here remaining horizontal as when first deposited, there upheaved or depressed, fractured or crushed, inclined or contorted; denuded by rain and rivers with the assistance of heat and cold, of frost and ice, in an unceasing series of changes, so that however varied the surface may be, with hill and dale, plains and uplands, mountain ranges and deep intervening valleys, these are as nothing to the diversities of interior structure, as exhibited in the sides of every alpine valley or precipitous escarpment, and made known to us by the work of the miner and the well-borer in every part of the world.

Structural Straight Lines on the Earth.

The great characteristic of the earth, both on its surface and in its interior, is thus seen to be extreme diversity both of form and structure, and this is further intensified by the varied texture, constitution, hardness, and density of the various rocks and debris of which it is composed. It is therefore not surprising that, with such a complex outer crust, we should nowhere find examples of those geometrical forms and almost world-wide straight lines that give such a remarkable, and as Mr. Lowell maintains, 'non-natural' character to the surface of Mars, but which, as it seems to me, of themselves afford prima facie evidence of a corresponding simplicity and uniformity in its internal structure.

Yet we are not ourselves by any means devoid of 'straight lines' structurally produced, in spite of every obstacle of diversity of form and texture, of softness and hardness, of lamination or crystallisation, which are adverse to such developments. Examples of these are the numerous 'faults' which occur in the harder rocks, and which often extend for great distances in almost perfect straight lines. In our own country we have the Tyneside and Craven faults in the North of England, which are 30 miles long and often 20 yards wide; but even more striking is the great Cleveland Dyke--a wall of volcanic rock dipping slightly towards the south, but sometimes being almost vertical, and stretching across the country, over hill and dale, in an almost perfect straight line from a point on the coast ten miles north of Scarborough, in a west-by-north direction, passing about two miles south of Stockton and terminating about six miles north-by-east of Barnard Castle, a distance of very nearly 60 miles. The great fault between the Highlands and Lowlands of Scotland extends across the country from Stonehaven to near Helensburgh, a distance of 120 miles; and there are very many more of less importance.

Much more extensive are some of the great continental dislocations, often forming valleys of considerable width and length. The Upper Rhine flows in one of these great valleys of subsidence for about 180 miles, from Mulhausen to Frankfort, in a generally straight line, though modified by denudation. Vaster still is the valley of the Jordan through the Sea of Galilee to the Dead Sea, continued by the Wady Arabah to the Gulf of Akaba, believed to form one vast geological depression or fracture extending in a straight line for 400 miles.

Thousands of such faults, dykes, or depressions exist in every part of the world, all believed to be due to the gradual shrinking of the heated interior to which the solid crust has to accommodate itself, and they are especially interesting and instructive for our present purpose as showing the tendency of such fractures of solid rock-material to extend to great lengths in straight lines, notwithstanding the extreme irregularity both in the surface contour as well as in the internal structures of the varied deposits and formations through which they pass.

Probable Origin of the Surface-features of Mars.

Returning now to Mars, let us consider the probable course of events from the point at which we left it. The heat produced by impact and condensation would be likely to release gases which had been in combination with some of the solid matter, or perhaps been itself in a solid state due to intense cold, and these, escaping outwards to the surface, would produce on a small scale a certain amount of upheaval and volcanic disturbance; and as an outer crust rapidly formed, a number of vents might remain as craters or craterlets in a moderate state of activity. Owing to the comparatively small force of gravity, the outer crust would become scoriaceous and more or less permeated by the gases, which would continue to escape through it, and this would facilitate the cooling of the whole of the heated outer crust, and allow it to become rather densely compacted. When the greater portion of the gases had thus escaped to the outer surface and assisted to form a scanty atmosphere, such as now exists, there would be no more internal disturbance and the cooling of the heated outer coating would steadily progress, resulting at last in a slightly heated, and later in a cold layer of moderate thickness and great general uniformity. Owing to the absence of rain and rivers, denudation such as we experience would be unknown, though the superficial scoriaceous crust might be partially broken up by expansion and contraction, and suffer a certain amount of atmospheric erosion.

The final result of this mode of aggregation would be, that the planet would consist of an outer layer of moderate thickness as compared with the central mass, which outer layer would have cooled from a highly heated state to a temperature considerably below the freezing-point, and this would have been all the time contracting upon a previously cold, and therefore non-contracting nucleus. The result would be that very early in the process great superficial tensions would be produced, which could only be relieved by cracks or fissures, which would initiate at points of weakness--probably at the craterlets already referred to--from which they would radiate in several directions. Each crack thus formed near the surface would, as cooling progressed, develop in length and depth; and owing to the general uniformity of the material, and possibly some amount of crystalline structure due to slow and continuous cooling down to a very low temperature, the cracks would tend to run on in straight lines and to extend vertically downwards, which two circumstances would necessarily result in their forming portions of 'great circles' on the planet's surface--the two great facts which Mr. Lowell appeals to as being especially 'non-natural.'

Symmetry of Basaltic Columns.

We have however one quite natural fact on our earth which serves to illustrate one of these two features, the direction of the downward fissure. This is, the comparatively common phenomenon of basaltic columns and 'Giant's Causeways.' The wonderful regularity of these, and especially the not unfrequent upright pillars in serried ranks, as in the palisades of the Hudson river, must have always impressed observers with their appearance of artificiality. Yet they are undoubtedly the result of the very slow cooling and contraction of melted rocks under compression by strata below and above them, so that, when once solidified, the mass was held in position and the tension produced by contraction could only be relieved by numerous very small cracks at short distances from each other in every direction, resulting in five, six, or seven-sided polygons, with sides only a few inches long. This contraction began of course at the coolest surface, generally the upper one; and observation of these columns in various positions has established the rule that their direction lengthways is always at right angles to the cooling surface, and thus, whenever this surface was horizontal, the columns became almost exactly vertical.

How this applies to Mars.

One of the features of the surface of Mars that Mr. Lowell describes with much confidence is, that it is wonderfully uniform and level, which of course it would be if it had once been in a liquid or plastic state, and not much disturbed since by volcanic or other internal movements. The result would be that cracks formed by contraction of the hardened outer crust would be vertical; and, in a generally uniform material at a very uniform temperature, these cracks would continue almost indefinitely in straight lines. The hardened and contracting surface being free to move laterally on account of there being a more heated and plastic layer below it, the cracks once initiated above would continually widen at the surface as they penetrated deeper and deeper into the slightly heated substratum. Now, as basalt begins to soften at about 1400?F. and the surface of Mars has cooled to at least the freezing-point--perhaps very much below it--the contraction would be so great that if the fissures produced were 500 miles apart they might be three miles wide at the surface, and, if only 100 miles apart, then about two-thirds of a mile wide.[15] But as the production of the fissures might have occupied perhaps millions of years, a considerable amount of atmospheric denudation would result, however slowly it acted. Expansion and contraction would wear away the edges and sides of the fissures, fill up many of them with the debris, and widen them at the surfaces to perhaps double their original size.[16]

[Footnote 15: The coefficient of contraction of basalt is 0.000006 for 1?F., which would lead to the results given here.]

[Footnote 16: Mr. W.H. Pickering observed clouds on Mars 15 miles high; these are the 'projections' seen on the terminator when the planet is partially illuminated. They were at first thought to be mountains; but during the opposition of 1894, more than 400 of them were seen at Flagstaff during nine months' observation. Usually they are of rare occurrence. They are seen to change in form and position from day to day, and Mr. Lowell is strongly of opinion that they are dust-storms, not what we term clouds. They were mostly about 13 miles high, indicating considerable aerial disturbance on the planet, and therefore capable of producing proportional surface denudation.]

Suggested Explanation of the 'Oases.'

The numerous round dots seen upon the 'canals,' and especially at points from which several canals radiate and where they intersect--termed 'oases' by Mr. Lowell and 'craterlets' by Mr. Pickering may be explained in two ways. Those from which several canals radiate may be true craters from which the gases imprisoned in the heated surface layers have gradually escaped. They would be situated at points of weakness in the crust, and become centres from which cracks would start during contraction. Those dots which occur at the crossing of two straight canals or cracks may have originated from the fact that at such intersections there would be four sharply-projecting angles, which, being exposed to the influence of alternate heat and cold (during day and night) on the two opposite surfaces, would inevitably in time become fractured and crumbled away, resulting in the formation of a roughly circular chasm which would become partly filled up by the debris. Those formed by cracks radiating from craterlets would also be subject to the same process of rounding off to an even greater extent; and thus would be produced the 'oases' of various sizes up to 50 miles or more in diameter recorded by Mr. Lowell and other observers.

Probable Function of the Great Fissures.

Mr. Pickering, as we have seen, supposes that these fissures give out the gases which, overflowing on each side, favour the growth of the supposed vegetation which renders the course of the canals visible, and this no doubt may have been the case during the remote periods when these cracks gave access to the heated portions of the surface layer. But it seems more probable that Mars has now cooled down to the almost uniform mean temperature it derives from solar heat, and that the fissures--now for the most part broad shallow valleys--serve merely as channels along which the liquids and heavy gases derived from the melting of the polar snows naturally flow, and, owing to their nearly level surfaces, overflow to a certain distance on each side of them.

Suggested Origin of the Blue Patches.

These heavy gases, mainly perhaps, as has been often suggested, carbon-dioxide, would, when in large quantity and of considerable depth, reflect a good deal of light, and, being almost inevitably dust-laden, might produce that blue tinge adjacent to the melting snow-caps which Mr. Lowell has erroneously assumed to be itself a proof of the presence of liquid water. Just as the blue of our sky is undoubtedly due to reflection from the ultra-minute dust particles in our higher atmosphere, similar particles brought down by the 'snow' from the higher Martian atmosphere might produce the blue tinge in the great volumes of heavy gas produced by its evaporation or liquefaction.

It may be noted that Mr. Lowell objects to the carbon-dioxide theory of the formation of the snow-caps, that this gas at low pressures does not liquefy, but passes at once from the solid to the gaseous state, and that only water remains liquid sufficiently long to produce the blue colour' which plays so large a part in his argument for the mild climate essential for an inhabited planet. But this argument, as I have already shown, is valueless. For only very deep water can possibly show a blue colour by reflected light, while a dust-laden atmosphere--especially with a layer of very dense gas at the bottom of it, as would be the case with the newly evaporated carbon-dioxide from the diminishing snow-cap --would provide the very conditions likely to produce this blue tinge of colour.

It may be considered a support to this view that carbonic-acid gas becomes liquid at--140?F. and solid at--162?F., temperatures far higher than we should expect to prevail in the polar and north temperate regions of Mars during a considerable part of the year, but such as might be reached there during the summer solstice when the `snows' so rapidly disappear, to be re-formed a few months later.

The Double Canals.

The curious phenomena of the 'double canals' are undoubtedly the most difficult to explain satisfactorily on any theory that has yet been suggested. They vary in distance apart from about 100 to 400 miles. In many cases they appear perfectly parallel, and Mr. Lowell gives us the impression that they are almost always so. But his maps show, in some cases, decided differences of width at the two extremities, indicating considerable want of parallelism. A few of the curved canals are also double.

There is one drawing in Mr. Lowell's book (p. 219) of the mouths, or starting points, of the Euphrates and Phison, two widely separated double canals diverging at an angle of about 40?from the same two oases, so that the two inner canals cross each other. Now this suggests two wide bands of weakness in the planet's crust radiating probably from within the dark tract called the 'Mare Icarium,' and that some widespread volcanic outburst initiated diverging cracks on either side of these bands. Something of this kind may have been the cause of most of the double canals, or they may have been started from two or more craterlets not far apart, the direction being at first decided by some local peculiarity of structure; and where begun continuing in straight lines owing to homogeneity or uniform density of material. This is very vague, but the phenomena are so remarkable, and so very imperfectly known at present, that nothing but suggestion can be attempted.

Concluding Remarks on the 'Canals.'

In this somewhat detailed exposition of a possible, and, I hope, a probable explanation of the surface-features of Mars, I have endeavoured to be guided by known facts or accepted theories both astronomical and geological. I think I may claim to have shown that there are some analogous features of terrestrial rock-structure to serve as guides towards a natural and intelligible explanation of the strange geometric markings discovered during the last thirty years, and which have raised this planet from comparative obscurity into a position of the very first rank both in astronomical and popular interest.

This wide-spread interest is very largely due to Mr. Lowell's devotion to its study, both in seeking out so admirable a position as regards altitude and climate, and in establishing there a first-class observatory; and also in bringing his discoveries before the public in connection with a theory so startling as to compel attention. I venture to think that his merit as one of our first astronomical observers will in no way be diminished by the rejection of his theory, and the substitution of one more in accordance with the actually observed facts.

APPENDIX.

A Suggested Experiment to Illustrate the 'Canals' of Mars.

If my explanation of the 'canals' should be substantially correct--that is, if they were produced by the contraction of a heated outward crust upon a cold, and therefore non-contracting interior, the result of such a condition might be shown experimentally.

Several baked clay balls might be formed to serve as cores, say of 8 to 10 inches in diameter. These being fixed within moulds of say half an inch to an inch greater diameter, the outer layer would be formed by pouring in some suitable heated liquid material, and releasing it from the mould as soon as consolidation occurs, so that it may cool rapidly from the outside. Some kinds of impure glass, or the brittle metals bismuth or antimony or alloys of these might be used, in order to see what form the resulting fractures would take. It would be well to have several duplicates of each ball, and, as soon as tension through contraction manifests itself, to try the effect of firing very small charges of small shot to ascertain whether such impacts would start radiating fractures. When taken from the moulds, the balls should be suspended in a slight current of air, and kept rotating, to reproduce the planetary condition as nearly as possible.

The exact size and material of the cores, the thickness of the heated outer crust, the material best suited to show fracture by contraction, and the details of their treatment, might be modified in various ways as suggested by the results first obtained. Such a series of experiments would probably throw further light on the physical conditions which have produced the gigantic system of fissures or channels we see upon the surface of Mars, though it would not, of course, prove that such conditions actually existed there. In such a speculative matter we can only be guided by probabilities, based upon whatever evidence is available.

CHAPTER VIII.

SUMMARY AND CONCLUSION.

This little volume has necessarily touched upon a great variety of subjects, in order to deal in a tolerably complete manner with the very extraordinary theories by which Mr. Lowell attempts to explain the unique features of the surface of the planet, which, by long-continued study, he has almost made his own. It may therefore be well to sum up the main points of the arguments against his view, introducing a few other facts and considerations which greatly strengthen my argument.

The one great feature of Mars which led Mr. Lowell to adopt the view of its being inhabited by a race of highly intelligent beings, and, with ever-increasing discovery to uphold this theory to the present time, is undoubtedly that of the so-called 'canals'--their straightness, their enormous length, their great abundance, and their extension over the planet's whole surface from one polar snow-cap to the other. The very immensity of this system, and its constant growth and extension during fifteen years of persistent observation, have so completely taken possession of his mind, that, after a very hasty glance at analogous facts and possibilities, he has declared them to be 'non-natural'-- therefore to be works of art--therefore to necessitate the presence of highly intelligent beings who have designed and constructed them. This idea has coloured or governed all his writings on the subject. The innumerable difficulties which it raises have been either ignored, or brushed aside on the flimsiest evidence. As examples, he never even discusses the totally inadequate water-supply for such worldwide irrigation, or the extreme irrationality of constructing so vast a canal-system the waste from which, by evaporation, when exposed to such desert conditions as he himself describes, would use up ten times the probable supply.

Again, he urges the 'purpose' displayed in these 'canals.' Their being all so straight, all describing great circles of the 'sphere,' all being so evidently arranged (as he thinks) either to carry water to some 'oasis' 2000 miles away, or to reach some arid region far over the equator in the opposite hemisphere! But he never considers the difficulties this implies. Everywhere these canals run for thousands of miles across waterless deserts, forming a system and indicating a purpose, the wonderful perfection of which he is never tired of dwelling upon (but which I myself can nowhere perceive).

Yet he never even attempts to explain how the Martians could have lived before this great system was planned and executed, or why they did not first utilise and render fertile the belt of land adjacent to the limits of the polar snows--why the method of irrigation did not, as with all human arts, begin gradually, at home, with terraces and channels to irrigate the land close to the source of the water. How, with such a desert as he describes three-fourths of Mars to be, did the inhabitants ever get to know anything of the equatorial regions and its needs, so as to start right away to supply those needs? All this, to my mind, is quite opposed to the idea of their being works of art, and altogether in favour of their being natural features of a globe as peculiar in origin and internal structure as it is in its surface-features. The explanation I have given, though of course hypothetical, is founded on known cosmical and terrestrial facts, and is, I suggest, far more scientific as well as more satisfactory than Mr. Lowell's wholly unsupported speculation. This view I have explained in some detail in the preceding chapter.

Mr. Lowell never even refers to the important question of loss by evaporation in these enormous open canals, or considers the undoubted fact that the only intelligent and practical way to convey a limited quantity of water such great distances would be by a system of water-tight and air-tight tubes laid under the ground. The mere attempt to use open canals for such a purpose shows complete ignorance and stupidity in these alleged very superior beings; while it is certain that, long before half of them were completed their failure to be of any use would have led any rational beings to cease constructing them.

He also fails to consider the difficulty, that, if these canals are necessary for existence in Mars, how did the inhabitants ever reach a sufficiently large population with surplus food and leisure enabling them to rise from the low condition of savages to one of civilisation, and ultimately to scientific knowledge? Here again is a dilemma which is hard to overcome. Only a dense population with ample means of subsistence could possibly have constructed such gigantic works; but, given these two conditions, no adequate motive existed for the conception and execution of them--even if they were likely to be of any use, which I have shown they could not be.

Further Considerations on the Climate of Mars.

Recurring now to the question of climate, which is all-important, Mr. Lowell never even discusses the essential point--the temperature that must necessarily result from an atmospheric envelope one-twelfth (or at most one-seventh) the density of our own; in either case corresponding to an altitude far greater than that of our highest mountains.[17] Surely this phenomenon, everywhere manifested on the earth even under the equator, of a regular decrease of temperature with altitude, the only cause of which is a less dense atmosphere, should have been fairly grappled with, and some attempt made to show why it should not apply to Mars, except the weak remark that on a level surface it will not have the same effect as on exposed mountain heights. But it does have the same effect, or very nearly so, on our lofty plateaux often hundreds of miles in extent, in proportion to their altitude. Quito, at 9350 ft. above the sea, has a mean temperature of about 57?F., giving a lowering of 23?from that of Manaos at the mouth of the Rio Negro. This is about a degree for each 400 feet, while the general fall for isolated mountains is about one degree in 340 feet according to Humboldt, who notes the above difference between the rate of cooling for altitude of the plains--or more usually sheltered valleys in which the towns are situated--and the exposed mountain sides. It will be seen that this lower rate would bring the temperature of Mars at the equator down to 20?F. below the freezing point of water from this cause alone.

[Footnote 17: A four inches barometer is equivalent to a height of 40,000 feet above sea-level with us.]

But all enquirers have admitted, that if conditions as to atmosphere were the same as on the earth, its greater distance from the sun would reduce the temperature to-31 F., equal to 63 below the freezing point. It is therefore certain that the combined effect of both causes must bring the temperature of Mars down to at least 70 or 80 below the freezing point.

The cause of this absolute dependence of terrestrial temperatures upon density of the air-envelope is seldom discussed in text-books either of geography or of physics, and there seems to be still some uncertainty about it. Some impute it wholly to the thinner air being unable to absorb and retain so much heat as that which is more dense; but if this were the case the soil at great altitudes not having so much of its heat taken up by the air should be warmer than below, since it undoubtedly receives more heat owing to the greater transparency of the air above it; but it certainly does not become warmer. The more correct view seems to be that the loss of heat by radiation is increased so much through the rarity of the air above it as to more than counterbalance the increased insolation, so that though the surface of the earth at a given altitude may receive 10 per cent. more direct sun-heat it loses by direct radiation, combined with diminished air and cloud-radiation, perhaps 20 or 25 per cent. more, whence there is a resultant cooling effect of 10 or 15 per cent. This acts by day as well as by night, so that the greater heat received at high altitudes does not warm the soil so much as a less amount of heat with a denser atmosphere.

This effect is further intensified by the fact that a less dense cannot absorb and transmit so much heat as a more dense atmosphere. Here then we have an absolute law of nature to be observed operating everywhere on the earth, and the mode of action of which is fairly well understood. This law is, that reduced atmospheric pressure increases radiation, or loss of heat, more rapidly than it increases insolation or gain of heat, so that the result is always a considerable lowering of temperature. What this lowering is can be seen in the universal fact, that even within the tropics perpetual snow covers the higher mountain summits, while on the high plains of the Andes, at 15,000 or 16,000 feet altitude, where there is very little or no snow, travellers are often frozen to death when delayed by storms; yet at this elevation the atmosphere has much more than double the density of that of Mars!

The error in Mr. Lowell's argument is, that he claims for the scanty atmosphere of Mars that it allows more sun-heat to reach the surface; but he omits to take account of the enormously increased loss of heat by direct radiation, as well as by the diminution of air-radiation, which together necessarily produce a great reduction of temperature.

It is this great principle of the prepotency of radiation over absorption with a diminishing atmosphere that explains the excessively low temperature of the moon's surface, a fact which also serves to indicate a very low temperature for Mars, as I have shown in Chapter VI. These two independent arguments-- from alpine temperatures and from those of the moon--support and enforce each other, and afford a conclusive proof (as against anything advanced by Mr. Lowell) that the temperature of Mars must be far too low to support animal life.

A third independent argument leading to the same result is Dr. Johnstone Stoney's proof that aqueous vapour cannot exist on Mars; and this fact Mr. Lowell does not attempt to controvert.

To put the whole case in the fewest possible words:

All physicists are agreed that, owing to the distance of Mars from the sun, it would have a mean temperature of about-35?F. (= 456?F. abs.) even if it had an atmosphere as dense as ours.

(2) But the very low temperatures on the earth under the equator, at a height where the barometer stands at about three times as high as on Mars, proves, that from scantiness of atmosphere alone Mars cannot possibly have a temperature as high as the freezing point of water; and this proof is supported by Langley's determination of the low maximum temperature of the full moon.

The combination of these two results must bring down the temperature of

Mars to a degree wholly incompatible with the existence of animal life.

(3) The quite independent proof that water-vapour cannot exist on Mars, and that therefore, the first essential of organic life--water--is non-existent.

The conclusion from these three independent proofs, which enforce each other in the multiple ratio of their respective weights, is therefore irresistible--that animal life, especially in its higher forms, cannot exist on the planet.

Mars, therefore, is not only uninhabited by intelligent beings such as Mr. Lowell postulates, but is absolutely UNINHABITABLE.

###